Edward L. Wolf
Nanophysics and Nanotechnology

Edward L. Wolf

Nanophysics and Nanotechnology

An Introduction to Modern Concepts
in Nanoscience

WILEY-VCH

WILEY-VCH Verlag GmbH & Co. KGaA

Author

Prof. Dr. Edward L. Wolf
Othmer Department of Chemical & Biological
Sciences & Engineering,
Polytechnic University Brooklyn
"Edward L. Wolf" ewolf@duke.poly.edu

Cover Picture
Assembling a ring of 48 Fe atoms on a (111) Cu
surface with an STM. The diameter of the ring is
14.3 nm

Library of Congress Card No.:
applied for

British Library Cataloguing-in-Publication Data
A catalogue record for this book is available from the
British Library.

Bibliographic information published by
Die Deutsche Bibliothek
Die Deutsche Bibliothek lists this publication in the
Deutsche Nationalbibliografie; detailed bibliographic
data is available in the Internet at
<http://dnb.ddb.de>.

Printed in the Federal Republic of Germany.

Printed on acid-free paper.

Typesetting   Kühn & Weyh, Satz und Medien,
Freiburg
Printing   Strauss GmbH, Mörlenbach
Bookbinding   Litges & Dopf Buchbinderei GmbH,
Heppenheim

ISBN   3-527-40407-4

To

Carol, Doug, Dave, Ben

And

Phill, Ned, Dan, Mehdi, Michael

Preface

This book originated with an elective sequence of two upper level undergraduate Physics courses, which I initiated at Polytechnic University. "Concepts of Nanotechnology" and "Techniques and Applications of Nanotechnology" are taken in the spring of the junior year and the following fall, and the students have a number of such sequences to choose from. I have been pleased with the quality, diversity (of major discipline), interest, and enthusiasm of the students who have taken the "Nano" sequence of courses, now midway in the second cycle of offering. Electrical engineering, computer engineering, computer science, mechanical engineering and chemical engineering are typical majors for these students, which facilitates breaking the class into interdisciplinary working groups who then prepare term papers and presentations that explore more deeply topics of their choice within the wealth of interesting topics in the area of nanotechnology. The Physics prerequisite for the course is 8 hours of calculus-based physics. The students have also had introductory Chemistry and an exposure to undergraduate mathematics and computer science.

I am grateful to my colleagues in the Interdisciplinary Physics Group for helping formulate the course, and in particular to Lorcan Folan and Harold Sjursen for help in getting the course approved for the undergraduate curriculum. Iwao Teraoka suggested, since I told him I had trouble finding a suitable textbook, that I should write such a book, and then introduced me to Ed Immergut, a wise and experienced consulting editor, who in turn helped me transform the course outlines into a book proposal. I am grateful to Rajinder Khosla for useful suggestions on the outline of the book. At Wiley-VCH I have benefited from the advice and technical support of Vera Palmer, Ron Schultz, Ulrike Werner and Anja Tschortner. At Polytechnic I have also been helped by DeShane Lyew and appreciate discussions and support from Stephen Arnold and Jovan Mijovic. My wife Carol has been a constant help in this project.

I hope this modest book, in addition to use as a textbook at the upper undergraduate or masters level, may more broadly be of interest to professionals who have had a basic background in physics and related subjects, and who have an interest in the developing fields of nanoscience and nanotechnology. I hope the book may play a career enhancing role for some readers. I have included some exercises to go with each chapter, and have also set off some tutorial material in half-tone sections of text, which many readers can pass over.

Nanophysics and Nanotechnology: An Introduction to Modern Concepts in Nanoscience. Edward L. Wolf
Copyright © 2004 WILEY-VCH Verlag GmbH & Co. KGaA, Weinheim
ISBN: 3-527-40407-4

At the beginning of the 21st century, with a wealth of knowledge in scientific and engineering disciplines, and really rapid ongoing advances, especially in areas of nanotechnology, robotics, and biotechnology, there may be a need also to look more broadly at the capabilities, opportunities, and possible pitfalls thus enabled. If there is to be a "posthuman era", a wide awareness of issues will doubtless be beneficial in making the best of it.

Edward L. Wolf
New York
July, 2004

Contents

Nanophysics and Nanotechnology: An Introduction to Modern Concepts in Nanoscience. Edward L. Wolf
Copyright © 2004 WILEY-VCH Verlag GmbH & Co. KGaA, Weinheim
ISBN: 3-527-40407-4

1
Introduction

Technology has to do with the application of scientific knowledge to the economic (profitable) production of goods and services. This book is concerned with the size or scale of working machines and devices in different forms of technology. It is particularly concerned with the smallest devices that are possible, and equally with the appropriate laws of nanometer-scale physics: "nanophysics", which are available to accurately predict behavior of matter on this invisible scale. Physical behavior at the nanometer scale is predicted accurately by quantum mechanics, represented by Schrodinger's equation. Schrodinger's equation provides a quantitative understanding of the structure and properties of atoms. Chemical matter, molecules, and even the cells of biology, being made of atoms, are therefore, in principle, accurately described (given enough computing power) by this well tested formulation of nanophysics.

There are often advantages in making devices smaller, as in modern semiconductor electronics. What are the limits to miniaturization, how small a device can be made? Any device must be composed of atoms, whose sizes are the order of 0.1 nanometer. Here the word "nanotechnology" will be associated with human-designed working devices in which some essential element or elements, produced in a controlled fashion, have sizes of 0.1 nm to thousands of nanometers, or, one Angstrom to one micron. There is thus an overlap with "microtechnology" at the micrometer size scale. Microelectronics is the most advanced present technology, apart from biology, whose complex operating units are on a scale as small as micrometers.

Although the literature of nanotechnology may refer to nanoscale machines, even "self-replicating machines built at the atomic level" [1], it is admitted that an "assembler breakthrough" [2] will be required for this to happen, and no nanoscale machines presently exist. In fact, scarcely any micrometer μm scale machines exist either, and it seems that the smallest mechanical machines readily available in a wide variety of forms are really on the millimeter scale, as in conventional wristwatches. (To avoid confusion, note that the prefix "micro" is sometimes applied, but never in this book, to larger scale techniques guided by optical microscopy, such as "microsurgery".)

The reader may correctly infer that Nanotechnology is presently more concept than fact, although it is certainly a media and funding reality. That the concept has

Nanophysics and Nanotechnology: An Introduction to Modern Concepts in Nanoscience. Edward L. Wolf
Copyright © 2004 WILEY-VCH Verlag GmbH & Co. KGaA, Weinheim
ISBN: 3-527-40407-4

great potential for technology, is the message to read from the funding and media attention to this topic.

The idea of the limiting size scale of a miniaturized technology is fundamentally interesting for several reasons. As sizes approach the atomic scale, the relevant physical laws change from the classical to the quantum-mechanical laws of nanophysics. The changes in behavior from classical, to "mesoscopic", to atomic scale, are broadly understood in contemporary physics, but the details in specific cases are complex and need to be worked out. While the changes from classical physics to nanophysics may mean that some existing devices will fail, the same changes open up possibilities for new devices.

A primary interest in the concept of nanotechnology comes from its connections with biology. The smallest forms of life, bacteria, cells, and the active components of living cells of biology, have sizes in the nanometer range. In fact, it may turn out that the only possibility for a viable complex nanotechnology is that represented by biology. Certainly the present understanding of molecular biology has been seen as an existence proof for "nanotechnology" by its pioneers and enthusiasts. In molecular biology, the "self replicating machines at the atomic level" are guided by DNA, replicated by RNA, specific molecules are "assembled" by enzymes and cells are replete with molecular scale motors, of which kinesin is one example. Ion channels, which allow (or block) specific ions (e.g., potassium or calcium) to enter a cell through its lipid wall, seem to be exquisitely engineered molecular scale devices where distinct conformations of protein molecules define an open channel vs. a closed channel.

Biological sensors such as the rods and cones of the retina and the nanoscale magnets found in magnetotactic bacteria appear to operate at the quantum limit of sensitivity. Understanding the operation of these sensors doubtless requires application of nanophysics. One might say that Darwinian evolution, a matter of odds of survival, has mastered the laws of quantum nanophysics, which are famously probabilistic in their nature. Understanding the role of quantum nanophysics entailed in the molecular building blocks of nature may inform the design of man-made sensors, motors, and perhaps much more, with expected advances in experimental and engineering techniques for nanotechnology.

In the improbable event that engineering, in the traditional sense, of molecular scale machines becomes possible, the most optimistic observers note that these invisible machines could be engineered to match the size scale of the molecules of biology. Medical nanomachines might then be possible, which could be directed to correct defects in cells, to kill dangerous cells, such as cancer cells, or even, most fancifully, to repair cell damage present after thawing of biological tissue, frozen as a means of preservation [3].

This book is intended to provide a guide to the ideas and physical concepts that allow an understanding of the changes that occur as the size scale shrinks toward the atomic scale. Our point of view is that a general introduction to the concepts of nanophysics will add greatly to the ability of students and professionals whose undergraduate training has been in engineering or applied science, to contribute in the various areas of nanotechnology. The broadly applicable concepts of nanophysics are worth study, as they do not become obsolete with inevitable changes in the forefront of technology.

1.1
Nanometers, Micrometers, Millimeters

A nanometer, 10^{-9} m, is about ten times the size of the smallest atoms, such as hydrogen and carbon, while a micron is barely larger than the wavelength of visible light, thus invisible to the human eye. A millimeter, the size of a pinhead, is roughly the smallest size available in present day machines. The range of scales from millimeters to nanometers is one million, which is also about the range of scales in present day mechanical technology, from the largest skyscrapers to the smallest conventional mechanical machine parts. The vast opportunity for making new machines, spanning almost six orders of magnitude from 1 mm to 1 nm, is one take on Richard Feynman's famous statement [4]: "there is plenty of room at the bottom". If L is taken as a typical length, 0.1 nm for an atom, perhaps 2 m for a human, this scale range in L would be 2×10^{10}. If the same scale range were to apply to an area, 0.1 nm by 0.1 nm vs 2 m $\times$ 2 m, the scale range for area L^2 is 4×10^{20}. Since a volume L^3 is enclosed by sides L, we can see that the number of atoms of size 0.1 nm in a $(2 \text{ m})^3$ volume is about 8×10^{30}. Recalling that Avogadro's number $N_A = 6.022 \times 10^{23}$ is the number of atoms in a gram-mole, supposing that the atoms were ^{12}C, molar mass 12 g; then the mass enclosed in the $(2 \text{ m})^3$ volume would be 15.9×10^4 kg, corresponding to a density 1.99×10^4 kg/m^3 (19.9 g/cc). (This is about 20 times the density of water, and higher than the densities of elemental carbon in its diamond and graphitic forms (which have densities 3.51 and 2.25 g/cc, respectively) because the equivalent size L of a carbon atom in these elemental forms slightly exceeds 0.1 nm.)

A primary working tool of the nanotechnologist is facility in scaling the magnitudes of various properties of interest, as the length scale L shrinks, e.g., from 1 mm to 1 nm.

Clearly, the number of atoms in a device scales as L^3. If a transistor on a micron scale contains 10^{12} atoms, then on a nanometer scale, $L'/L = 10^{-3}$ it will contain 1000 atoms, likely too few to preserve its function!

Normally, we will think of scaling as an isotropic scale reduction in three dimensions. However, scaling can be thought of usefully when applied only to one or two dimensions, scaling a cube to a two-dimensional (2D) sheet of thickness a or to a one-dimensional (1D) tube or "nanowire" of cross sectional area a^2. The term "zero-dimensional" is used to describe an object small in all three dimensions, having volume a^3. In electronics, a zero-dimensional object (a nanometer sized cube a^3 of semiconductor) is called a "quantum dot" (QD) or "artificial atom" because its electron states are few, sharply separated in energy, and thus resemble the electronic states of an atom.

As we will see, a quantum dot also typically has so small a radius a, with correspondingly small electrical capacitance $C = 4\pi\varepsilon\varepsilon_o a$ (where $\varepsilon\varepsilon_o$ is the dielectric constant of the medium in which the QD is immersed), that the electrical charging energy $U = Q^2/2C$ is "large". (In many situations, a "large" energy is one that exceeds the thermal excitation energy, k_BT, for $T = 300$ K, basically room temperature. Here T is the absolute Kelvin temperature, and k_B is Boltzmann's constant, 1.38×10^{-23} J/K.) In this situation, a change in the charge Q on the QD by even one electron charge e, may effectively, by the "large" change in U, switch off the possibility of the QD being part of the path of flow for an external current.

This is the basic idea of the "single electron transistor". The role of the quantum dot or QD in this application resembles the role of the grid in the vacuum triode, but only one extra electron change of charge on the "grid" turns the device off. To make a device of this sort work at room temperature requires that the QD be tiny, only a few nm in size.

Plenty of room at the bottom

Think of reducing the scale of working devices and machines from 1mm to 1nm, six orders of magnitude! Over most of this scaling range, perhaps the first five orders of magnitude, down to 10 nm (100 Angstroms), the laws of classical Newtonian physics may well suffice to describe changes in behavior. This classical range of scaling is so large, and the changes in magnitudes of important physical properties, such as resonant frequencies, are so great, that completely different applications may appear.

Scaling the xylophone

The familiar xylophone produces musical sounds when its keys (a linear array of rectangular bars of dimensions $a \times b \times c$, with progressively longer key lengths c producing lower audio frequencies) are struck by a mallet and go into transverse vibration perpendicular to the smallest, a, dimension. The traditional "middle C" in music corresponds to 256 Hz. If the size scale of the xylophone key is reduced to the micrometer scale, as has recently been achieved, using the semiconductor technology, and the mallet is replaced by electromagnetic excitation, the same transverse mechanical oscillations occur, and are measured to approach the Gigahertz (10^9 Hz) range [5]!

The measured frequencies of the micrometer scale xylophone keys are still accurately described by the laws of classical physics. (Actually the oscillators that have been successfully miniaturized, see Figure 1.1, differ slightly from xylophone keys, in that they are clamped at both ends, rather than being loosely suspended. Very similar equations are known to apply in this case.) Oscillators whose frequencies approach the GHz range have completely different applications than those in the musical audio range!

Could such elements be used in new devices to replace Klystrons and Gunn oscillators, conventional sources of GHz radiation? If means could be found to fabricate "xylophone keys" scaling down from the micrometer range to the nanometer range, classical physics would presumably apply almost down to the molecular scale. The limiting vibration frequencies would be those of diatomic molecules, which lie in the range $10^{13} - 10^{14}$ Hz. For comparison, the frequency of light used in fiberoptic communication is about 2×10^{14} Hz.

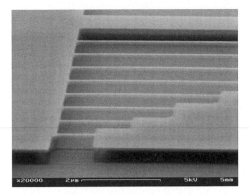

Figure 1.1 Silicon nanowires in a harp-like array. Due to the clamping of the single-crystal silicon bars at each end, and the lack of applied tension, the situation is more like an array of xylophone keys. The resonant frequency of the wire of 2 micrometer length is about 400 MHz. After Reference [5]

Reliability of concepts and approximate parameter values down to about $L = 10$ nm (100 atoms)

The large extent of the "classical" range of scaling, from 1 mm down to perhaps 10 nm, is related to the stability (constancy) of the basic microscopic properties of condensed matter (conventional building and engineering materials) almost down to the scale L of 10 nm or 100 atoms in line, or a million atoms per cube.

Typical microscopic properties of condensed matter are the interatomic spacing, the mass density, the bulk speed of sound v_s, Young's modulus Y, the bulk modulus B, the cohesive energy U_o, the electrical resistivity ρ, thermal conductivity K, the relative magnetic and dielectric susceptibilities κ and ε, the Fermi energy E_F and work function φ of a metal, and the bandgap of a semiconductor or insulator, E_g. A timely example in which bulk properties are retained down to nanometer sample sizes is afforded by the CdSe "quantum dot" fluorescent markers, which are described below.

Nanophysics built into the properties of bulk matter

Even if we can describe the size scale of 1 mm – 10 nm as one of "classical scaling", before distinctly size-related anomalies are strongly apparent, a nanotechnologist must appreciate that many properties of bulk condensed matter already require concepts of nanophysics for their understanding. This might seem obvious, in that atoms themselves are completely nanophysical in their structure and behavior!

Beyond this, however, the basic modern understanding of semiconductors, involving energy bands, forbidden gaps, and effective masses m^* for free electrons and free holes, is based on nanophysics in the form of Schrodinger's equation as applied to a periodic structure.

Periodicity, a repeated unit cell of dimensions a,b,c (in three dimensions) profoundly alters the way an electron (or a "hole", which is the inherently positively charged absence of an electron) moves in a solid. As we will discuss more complete-ly below, ranges (bands) of energy of the free carrier exist for which the carrier will pass through the periodic solid with no scattering at all, much in the same way that an electromagnetic wave will propagate without attenuation in the passband of a transmission line. In energy ranges between the allowed bands, gaps appear, where no moving carriers are possible, in analogy to the lack of signal transmission in the stopband frequency range of a transmission line.

So, the "classical" range of scaling as mentioned above is one in which the conse-quences of periodicity for the motions of electrons and holes (wildly "non-classical", if referred to Newton's Laws, for example) are unchanged. In practice, the properties of a regular array of 100 atoms on a side, a nanocrystal containing only a million atoms, is still large enough to be accurately described by the methods of solid state physics. If the material is crystalline, the properties of a sample of 10^6 atoms are likely to be an approximate guide to the properties of a bulk sample. To extrapolate the bulk properties from a 100-atom-per-side simulation may not be too far off.

It is probably clear that a basic understanding of the ideas, and also the fabrica-tion methods, of semiconductor physics is likely to be a useful tool for the scientist or engineer who will work in nanotechnology. Almost all devices in the Micro-elec-tromechanical Systems (MEMS) category, including accelerometers, related angular rotation sensors, and more, are presently fabricated using the semiconductor micro-technology.

The second, and more challenging question, for the nanotechnologist, is to under-stand and hopefully to exploit those changes in physical behavior that occur at the end of the classical scaling range. The "end of the scaling" is the size scale of atoms and molecules, where nanophysics is the proven conceptual replacement of the laws of classical physics. Modern physics, which includes quantum mechanics as a description of matter on a nanometer scale, is a fully developed and proven subject whose application to real situations is limited only by modeling and computational competence.

In the modern era, simulations and approximate solutions increasingly facilitate the application of nanophysics to almost any problem of interest. Many central prob-lems are already (adequately, or more than adequately) solved in the extensive litera-tures of theoretical chemistry, biophysics, condensed matter physics and semicon-ductor device physics. The practical problem is to find the relevant work, and, fre-quently, to convert the notation and units systems to apply the results to the prob-lem at hand.

It is worth saying that information has no inherent (i.e., zero) size. The density of information that can be stored is limited only by the coding element, be it a bead on an abacus, a magnetized region on a hard disk, a charge on a CMOS capacitor, a nanoscale indentation on a plastic recording surface, the presence or absence of a particular atom at a specified location, or the presence of an "up" or "down" electron-ic or nuclear spin (magnetic moment) on a density of atoms in condensed matter, $(0.1 \, \text{nm})^{-3} = 10^{30}/\text{m}^3 = 10^{24}/\text{cm}^3$. If these coding elements are on a surface, then the limiting density is $(0.1 \, \text{nm})^{-2} = 10^{20}/\text{m}^2$, or $6.45 \times 10^{16}/\text{in}^2$.

The principal limitation may be the physical size of the reading element, which historically would be a coil of wire (solenoid) in the case of the magnetic bit. The limiting density of information in the presently advancing technology of magnetic computer hard disk drives is about $100\,Gb/in^2$, or $10^{11}/in^2$. It appears that non-magnetic technologies, perhaps based on arrays of AFM tips writing onto a plastic film such as polymethylmethacrylate (PMMA), may eventually overtake the magnetic technology.

1.2
Moore's Law

The computer chip is certainly one of the preeminent accomplishments of 20th century technology, making vastly expanded computational speed available in smaller size and at lower cost. Computers and email communication are almost universally available in modern society. Perhaps the most revolutionary results of computer technology are the universal availability of email to the informed and at least minimally endowed, and magnificent search engines such as Google. Without an unexpected return to the past, which might roll back this major human progress it seems rationally that computers have ushered in a new era of information, connectedness, and enlightenment in human existence.

Moore's empirical law summarizes the "economy of scale" in getting the same function by making the working elements ever smaller. (It turns out, as we will see, that smaller means faster, characteristically enhancing the advantage in miniaturization). In the ancient abacus, bead positions represent binary numbers, with infor-

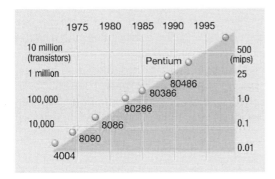

Figure 1.2 Moore's Law. [6]. The number of transistors in successive generations of computer chips has risen exponentially, doubling every 1.5 years or so. The notation "mips" on right ordinate is "million instructions per second". Gordon Moore, co-founder of Intel, Inc. predicted this growth pattern in 1965, when a silicon chip contained only 30 transistors! The number of Dynamic Random Access Memory (DRAM) cells follows a similar growth pattern. The growth is largely due to continuing reduction in the size of key elements in the devices, to about 100 nm, with improvements in optical photolithography. Clock speeds have similarly increased, presently around 2 GHz. For a summary, see [7]

mation recorded on a scale of perhaps 1 bit [(0,1) or (yes/no)] per cm^2. In silicon microelectronic technology an easily produced memory cell size of one micron corresponds to 10^{12} bits/cm^2 (one Tb/cm^2). Equally important is the continually reducing size of the magnetic disk memory element (and of the corresponding read/write sensor head) making possible the ~100 Gb disk memories of contemporary laptop computers. The continuing improvements in performance (reductions in size of the performing elements), empirically summarized by Moore's Law (a doubling of performance every 1.5 years, or so), arise from corresponding reductions in the size scale of the computer chip, aided by the advertising-related market demand.

The vast improvements from the abacus to the Pentium chip exemplify the promise of nanotechnology. Please note that this is all still in the range of "classical scaling"! The computer experts are absolutely sure that nanophysical effects are so far negligible.

The semiconductor industry, having produced a blockbuster performance over decades, transforming advanced society and suitably enriching its players and stockholders, is concerned about its next act!

The next act in the semiconductor industry, if a second act indeed shows up, must deal with the nanophysical rules. Any new technology, if such is feasible, will have to compete with a base of universally available applied computation, at unimaginably low costs. If Terahertz speed computers with 100 Mb randomly accessible memories and 100 Gb hard drives, indeed become a commodity, what can compete with that? Silicon technology is a hard act to follow.

Nanotechnology, taken literally, also represents the physically possible limit of such improvements. The limit of technology is also evident, since the smallest possible interconnecting wire on the chip must be at least 100 atoms across! Moore's law empirically has characterized the semiconductor industry's success in providing faster and faster computers of increasing sophistication and continually falling price. Success has been obtained with a larger number of transistors per chip made possible by finer and finer scales of the wiring and active components on the silicon chips. There is a challenge to the continuation of this trend (Moore's Law) from the economic reality of steeply increasing plant cost (to realize reduced linewidths and smaller transistors).

The fundamental challenge to the continuation of this trend (Moore's Law) from the change of physical behavior as the atomic size limit is approached, is a central topic in this book.

1.3
Esaki's Quantum Tunneling Diode

The tunneling effect is basic in quantum mechanics, a fundamental consequence of the probabilistic wave function as a measure of the location of a particle. Unlike a tennis ball, a tiny electron may penetrate a barrier. This effect was first exploited in semiconductor technology by Leo Esaki, who discovered that the current–voltage (I/V) curves of semiconductor p–n junction rectifier diodes (when the barrier was made very thin, by increasing the dopant concentrations) became anomalous, and in fact double-valued. The forward bias I vs. V plot, normally a rising exponential $\exp(eV/kT)$, was preceded by a distinct "current hump" starting at zero bias and extending to $V = 50$ mV or so. Between the region of the "hump" and the onset of the conventional exponential current rise there was a region of negative slope, $dI/dV < 0$!

The planar junction between an N-type region and a P-type region in a semiconductor such as Si contains a "depletion region" separating conductive regions filled with free electrons on the N-side and free holes on the P-side. It is a useful non-trivial exercise in semiconductor physics to show that the width W of the depletion region is

$$W = [2\varepsilon\varepsilon_o V_B(N_D + N_A)/e(N_D N_A)]^{1/2}. \tag{1.1}$$

Here $\varepsilon\varepsilon_o$ is the dielectric constant, e the electron charge, V_B is the energy shift in the bands across the junction, and N_D and N_A, respectively, are the concentrations of donor and acceptor impurities.

The change in electrical behavior (the negative resistance range) resulting from the electron tunneling (in the thin depletion region limit) made possible an entirely new effect, an oscillation, at an extremely high frequency! (As often happens with the continuing advance of technology, this pioneering device has been largely supplanted as an oscillator by the Gunn diode, which is easier to manufacture.)

The Esaki tunnel diode is perhaps the first example in which the appearance of quantum physics at the limit of a small size led to a new device. In our terminology the depletion layer tunneling barrier is two-dimensional, with only one small dimension, the depletion layer thickness W. The Esaki diode falls into our classification as an element of nanotechnology, since the controlled small barrier W is only a few nanometers in thickness.

1.4
Quantum Dots of Many Colors

"Quantum dots" (QDs) of CdSe and similar semiconductors are grown in carefully controlled solution precipitation with controlled sizes in the range $L = 4$ or 5 nm. It is found that the wavelength (color) of strong fluorescent light emitted by these quantum dots under ultraviolet (uv) light illumination depends sensitively on the size L.

There are enough atoms in this particle to effectively validate the concepts of solid state physics, which include electron bands, forbidden energy band gaps, electron and hole effective masses, and more.

Still, the particle is small enough to be called an "artificial atom", characterized by discrete sharp electron energy states, and discrete sharp absorption and emission wavelengths for photons.

Transmission electron microscope (TEM) images of such nanocrystals, which may contain only 50 000 atoms, reveal perfectly ordered crystals having the bulk

Figure 1.3 Transmission Electron Micrograph (TEM) Image of one 5 nm CdSe quantum dot particle, courtesy Andreas Kornowski, University of Hamburg, Germany

crystal structure and nearly the bulk lattice constant. Quantitative analysis of the light emission process in QDs suggests that the bandgap, effective masses of electrons and holes, and other microscopic material properties are very close to their values in large crystals of the same material. The light emission in all cases comes from radiative recombination of an electron and a hole, created initially by the shorter wavelength illumination.

The energy E_R released in the recombination is given entirely to a photon (the quantum unit of light), according to the relation $E_R = h\nu = hc/\lambda$. Here ν and λ are, respectively, the frequency and wavelength of the emitted light, c is the speed of light 3×10^8 m/s, and h is Planck's constant $h = 6.63 \times 10^{-34}$ Js $= 4.136 \times 10^{-15}$ eVs. The color of the emitted light is controlled by the choice of L, since $E_R = E_G + E_e + E_h$, where E_G is the semiconductor bandgap, and the electron and hole confinement energies, E_e and E_h, respectively, become larger with decreasing L.

It is an elementary exercise in nanophysics, which will be demonstrated in Chapter 4, to show that these confinement (blue-shift) energies are proportional to $1/L^2$. Since these terms increase the energy of the emitted photon, they act to shorten the wavelength of the light relative to that emitted by the bulk semiconductor, an effect referred to as the "blue shift" of light from the quantum dot.

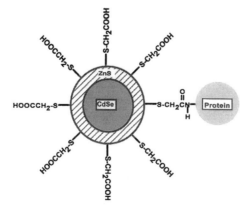

Figure 1.4 Schematic of quantum dot with coatings suitable to assure water solubility, for application in biological tissue. This ZnS-capped CdSe quantum dot is covalently coupled to a protein by mercaptoacetic acid. The typical QD core size is 4.2 nm. [8]

These nanocrystals are used in biological research as markers for particular kinds of cells, as observed under an optical microscope with background ultraviolet light (uv) illumination.

In these applications, the basic semiconductor QD crystal is typically coated with a thin layer to make it impervious to (and soluble in) an aqueous biological environment. A further coating may then be applied which allows the QD to preferentially bond to a specific biological cell or cell component of interest. Such a coated quantum dot is shown in Figure 1.4 [8]. These authors say that the quantum dots they use as luminescent labels are 20 times as bright, 100 times as stable against photobleaching, and have emission spectra three times sharper than conventional organic dyes such as rhodamine.

The biological researcher may, for example, see the outer cell surface illuminated in green while the surface of the inner cell nucleus may be illuminated in red, all under illumination of the whole field with light of a single shorter wavelength.

1.5
GMR 40 Gb Hard Drive "Read" Heads

In modern computers, the hard disk encodes information in the form of a linear array of planar ferromagnetic regions, or bits. The performance has recently improved with the discovery of the Giant Magnetoresistance (GMR) effect allowing a smaller assembly (read head), to scan the magnetic data. The ferromagnetic bits are written into (and read from) the disk surface, which is uniformly coated with a ferromagnetic film having a small coercive field. This is a "soft" ferromagnet, so that a small imposed magnetic field B can easily establish the ferromagnetic magnetization M along the direction of the applied B field. Both writing and reading operations are accomplished by the "read head".

The density of information that can be stored in a magnetic disk is fundamentally limited by the minimum size of a ferromagnetic "domain". Ferromagnetism is a cooperative nanophysical effect requiring a minimum number of atoms: below this number the individual atomic magnetic moments remain independent of each other. It is estimated that this "super-paramagnetic limit" is on the order of 100 Gb/in^2.

The practical limit, however, has historically been the size of the "read head", as sketched in Figure 1.5, which on the one hand impresses a local magnetic field **B** on the local surface region to create the magnetized domain, and then also senses the magnetic field of the magnetic domain so produced. In present technology, the linear bits are about 100 nm in length (**M** along the track) and have widths in the range 0.3 – 1.0 μm. The ferromagnetic domain magnetization **M** is parallel or anti-parallel to the linear track.

The localized, perpendicular, magnetic fields **B** that appear at the junctions between parallel and anti-parallel bits are sensed by the read head. The width of the transition region between adjacent bits, in which the localized magnetic field is present, is between 10 and 100 nm. The localized **B** fields extend linearly across the track and point upward (or down) from the disk surface, as shown in Figure 1.5.

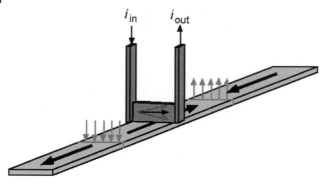

i_{in} i_{out}

Figure 1.5 Schematic diagram of the GMR read head, showing
two current leads connected by the sensing element, which itself
is a conducting copper sheet sandwiched between a hard and a
soft magnet. [9]

The state-of-the-art magnetic field sensor is an exquisitely thin sandwich of magnetic and non-magnetic metals oriented vertically to intercept the fringe **B** field between adjacent bits. The total thickness of the sensor sandwich, along the direction
of the track, is presently about 80 nm, but this thickness may soon fall to 20 nm.
The GMR detector sandwich is comprised of a sensing soft ferromagnetic layer of
NiFe alloy, a Cu spacer, and a "magnetically hard" Co ferromagnetic film. A sensing
current is directed along the sandwich in the direction transverse to the track, and
the voltage across the sandwich, which is sensitive to the magnetic field in the plane
of the sandwich, is measured. In this read-head sandwich the Cu layer is about
15 atoms in thickness! The sensitivity of this GMR magnetoresistive sensor is presently in the vicinity of 1% per Oersted.

Writing the magnetic bits is accomplished by an integrated component of the
"read head" (not shown in Figure 1.5) which generates a surface localized magnetic
field **B** parallel or antiparallel to the track. The local surface magnetic field is produced inductively in a closed planar loop, resembling an open-ended box, of thin
magnetic film, which is interrupted by a linear gap facing the disk, transverse to the
track.

These "read head" units are fabricated in mass, using methods of the silicon
lithographic microtechnology. It is expected that even smaller sensor devices and
higher storage densities may be possible with advances in the silicon fabrication
technology.

These units, which have had a large economic impact, are a demonstration of
nanotechnology in that their closely controlled dimensions are in the nm range. The
mechanism of greatly enhanced magnetic field sensitivity in the Giant Magnetoresistance Effect (GMR) is also fully nanophysical in nature, an example of (probably
unexpectedly) better results at the quantum limit of the scaling process.

1.6
Accelerometers in your Car

Modern cars have airbags which inflate in crashes to protect drivers and passengers from sudden accelerations. Micro-electro-mechanical semiconductor acceleration sensors (accelerometers) are located in bumpers which quickly inflate the airbags. The basic accelerometer is a mass m attached by a spring of constant k to the frame of the sensor device, itself secured to the automobile frame. If the car frame (and thus the frame of the sensor device) undergo a strong acceleration, the spring connecting the mass to the sensor frame will extend or contract, leading to a motion of the mass m relative to the frame of the sensor device. This deflection is measured, for example, by change in a capacitance, which then triggers expansion of the airbag. This microelectromechanical (MEM) device is mass-produced in an integrated package including relevant electronics, using the methods of silicon microelectronics.

Newton's laws of motion describe the position, x, the velocity, $v = dx/dt$, and the acceleration $a = d^2x/dt^2$ of a mass m which may be acted upon by a force F, according to

$$md^2x/dt^2 = F. \qquad (1.2)$$

Kinematics describes the relations among x, v, a, and t. As an example, the time-dependent position x under uniform acceleration a is

$$x = x_0 + v_0 t + at^2, \qquad (1.3)$$

where x_0 and v_0, respectively, are the position and velocity at $t = 0$. Also, if a time-varying acceleration $a(t)$ is known, and $x_0 = 0$ and $v_0 = 0$, then

$$x(t) = \iint a(t)dt^2. \qquad (1.4)$$

Newton's First Law states that in the absence of a force, the mass m remains at rest if initially so, and if initially in motion continues unchanged in that motion. (These laws are valid only if the coordinate system in which the observations are made is one of uniform motion, and certainly do not apply in an accelerated frame of reference such as a carrousel or a merry-go-round. For most purposes the earth's surface, although accelerated toward the earth's rotation axis, can be regarded as an "inertial frame of reference", i.e., Newton's Laws are useful.)

The Second Law is $F = ma$, (1.2).

The Third Law states that for two masses in contact, the force exerted by the first on the second is equal and opposite to the force exerted by the second on the first.

A more sophisticated version of such an accelerometer, arranged to record accelerations in x,y,z directions, and equipped with integrating electronics, can be used to record the three-dimensional displacement over time.

These devices are not presently built on a nanometer scale, of course, but are one example of a wide class of microelectronic sensors that could be made on smaller scales as semiconductor technology advances, and if smaller devices are useful.

1.7
Nanopore Filters

The original nanopore (Nuclepore) filters [10,11] are sheets of polycarbonate of $6 - 11\,\mu m$ thickness with closely spaced arrays of parallel holes running through the sheet. The filters are available with pore sizes rated from $0.015\,\mu m$ – $12.0\,\mu m$ ($15\,nm$ – $12\,000\,nm$). The holes are made by exposing the polycarbonate sheets to perpendicular flux of ionizing α particles, which produce linear paths of atomic scale damage in the polycarbonate. Controlled chemical etching is then employed to establish and enlarge the parallel holes to the desired diameter. This scheme is an example of nanotechnology.

The filters are robust and can have a very substantial throughput, with up to 12% of the area being open. The smallest filters will block passage of bacteria and perhaps even some viruses, and are used in many applications including water filters for hikers.

A second class of filters (Anapore) was later established, formed of alumina grown by anodic oxidation of aluminum metal. These filters are more porous, up to 40%, and are stronger and more temperature resistant than the polycarbonate filters. The Anapore filters have been used, for example, to produce dense arrays of nanowires. Nanowires are obtained using a hot press to force a ductile metal into the pores of the nanopore alumina filter.

1.8
Nanoscale Elements in Traditional Technologies

From the present knowledge of materials it is understood that the beautiful colors of stained glass windows originate in nanometer scale metal particles present in the glass. These metal particles have scattering resonances for light of specific wavelengths, depending on the particle size L. The particle size distribution, in turn, will depend upon the choice of metal impurity, its concentration, and the heat treatment of the glass. When the metallic particles in the glass are illuminated, they preferentially scatter light of particular colors. Neutral density filters marketed for photographic application also have distributions of small particles embedded in glass.

Carbon black, commonly known as soot, which contains nanometer-sized particles of carbon, was used very early as an additive to the rubber in automobile tires.

(As we now know, carbon black contains small amounts of ^{60}C (Buckminsterfuller-ene), other fullerenes, and graphitic nanotubes of various types.)

The AgBr and AgI crystals of conventional photography are nanometer-sized single crystals embedded in a thin gelatin matrix. It appears that the fundamental light absorption in these crystals is close to the quantum sensitivity limit. It further appears that the nanoscopic changes in these tiny crystals, which occur upon absorption of one or more light photons, enable them to be turned into larger metallic silver particles in the conventional photographic development process. The conventional photographic negative image is an array of solid silver grains embedded in a gelatin matrix. As such it is remarkably stable as a record, over decades or more.

The drugs that are so important in everyday life (and are also of huge economic importance), including caffeine, aspirin and many more, are specific molecules of nanometer size, typically containing fewer than 100 atoms.

Controlled precipitation chemistry for example is employed to produce uniform nanometer spheres of polystyrene, which have long been marketed as calibration markers for transmission electron microscopy.

References

[1] R. Kurzweil, *The Age of Spiritual Machines*, (Penguin Books, New York, 1999), page 140.

[2] K. E. Drexler, *Engines of Creation*, (Anchor Books, New York, 1986), page 49.

[3] K. E. Drexler, op. cit., p. 268.

[4] R. Fcynman, "There's plenty of room at the bottom", in *Miniaturization*, edited by H.D. Gilbert (Reinhold, New York, 1961).

[5] Reprinted with permission from D.W. Carr *et al.*, Appl. Phy. Lett. **75**, 920 (1999). Copyright 1999. American Institute of Physics.

[6] Reprinted with permission from Nature: P. Ball, Nature **406**, 118–120 (2000). Copyright 2000. Macmillan Publishers Ltd.

[7] M. Lundstrom, Science **299**, 210 (2003).

[8] Reprinted with permission from C.W. Warren & N. Shumig, Science **281**, 2016–2018 (1998). Copyright 1998 AAAS.

[9] Courtesy G.A. Prinz, U.S. Naval Research Laboratory, Washington, DC.

[10] These filters are manufactured by Nuclepore Corporation, 7035 Commerce Circle, Pleasanton, CA 94566.

[11] G.P.Crawford, L.M. Steele, R. Ondris-Crawford, G. S. Iannocchione, C. J. Yeager, J. W. Doane, and D. Finotello, J. Chem. Phys. **96**, 7788 (1992).

2

Systematics of Making Things Smaller, Pre-quantum

2.1
Mechanical Frequencies Increase in Small Systems

Mechanical resonance frequencies depend on the dimensions of the system at hand. For the simple pendulum, $\omega = (g/l)^{1/2}$, where l is the length of the pendulum rod and g is the acceleration of gravity. The period $T = 2\pi/\omega$ of the pendulum of a grandfather clock is exactly one second, and corresponds to a length l about one meter (depending on the exact local value of the gravitational acceleration, g, which is approximately $9.8 \, \text{m/s}^2$). The relation $\omega = (g/l)^{1/2}$ indicates that the period of the pendulum scales as $\sqrt{l}$, so that for l one micron T is one ms; a one-micron sized grandfather clock (oscillator) would generate a 1000 Hz tone. Either clock could be used (in conjunction with a separate frequency measurement device or counter), to measure the local vertical acceleration $a = d^2y/dt^2$ according to $\omega = [(g+a)/l]^{1/2}$.

If used in this way as an accelerometer, note that the miniature version has a much faster response time, 1ms, than the original grandfather clock, which would have a time resolution of about a second.

A mass m attached to a rigid support by a spring of constant k has a resonance frequency $\omega = (k/m)^{1/2}$. This oscillator, and the pendulum of the grandfather clock, are examples of simple harmonic oscillation (SHO).

Simple harmonic oscillation occurs when a displacement of a mass m in a given direction, x, produces an (oppositely directed) force $F = -kx$. The effective spring constant k has units of N/m in SI units. According to Newton's Second Law (1.2) $F = ma = md^2x/dt^2$. Applied to the mass on the spring, this gives the differential equation

$$d^2x/dt^2 + (k/m)x = 0. \tag{2.1}$$

$$x = x_{\text{max}} \cos(\omega t + \delta) \tag{2.2}$$

is a solution of the equation for arbitrary amplitude x_{max} and arbitrary initial phase angle δ, but only when

$$\omega = (k/m)^{1/2}. \tag{2.3}$$

Nanophysics and Nanotechnology: An Introduction to Modern Concepts in Nanoscience. Edward L. Wolf
Copyright © 2004 WILEY-VCH Verlag GmbH & Co. KGaA, Weinheim
ISBN: 3-527-40407-4

The period of the motion is therefore $T = 2\pi(m/k)^{1/2}$. The maximum values of the speed $v = dx/dt$ and the acceleration d^2x/dt^2 are seen to be $x_{max}\omega$ and $x_{max}\omega^2$, respectively. The total energy $E = U + K$ in the motion is constant and equal to $1/2kx_{max}^2$. (In nanophysics, which is needed when the mass m is on an atomic scale, the same frequency $\omega = (k/m)^{1/2}$ is found, but the energies are restricted to $E_n = (n + 1/2)\,\hbar\omega$, where the quantum number n can take zero or positive integer values, $\hbar = h/2\pi$, and Planck's constant is $h = 6.67 \times 10^{-34}$Joule·s.)

Simple harmonic oscillation is a more widely useful concept than one might think at first, because it is applicable to any system near a minimum, say x_o, in the system potential energy $U(x)$. Near x_o the potential energy $U(x)$ can be closely approximated as a constant plus $k(x-x_o)^2/2$, leading to the same resonance frequency for oscillations of amplitude A in $x-x_o$. An important example is in molecular bonding, where x_o is the interatomic spacing.

More generally, the behavior applies whenever the differential equation appears, and the resonant frequency will be the square root of the coefficient of x in the equation. In the case of the pendulum, if $x \cong L$ is the horizontal displacement of the mass m., then $F \cong -g\,x/L$ and $\omega = (g/L)^{1/2}$.

Considering the mass and spring to be three-dimensional, mass m will vary as L^3 and k will vary as L, leading to $\omega \propto \alpha\,L^{-1}$. Frequency inversely proportional to length scale is typical of mechanical oscillators such as a violin or piano string and the frequency generated by a solid rod of length L struck on the end. In these cases the period of the oscillation T is the time for the wave to travel $2L$, hence $T = 2L/v$. (This is the same as $L = \lambda/2$, where $\lambda = vT$ is the wavelength. If the boundary conditions are different at the two ends, as in a clarinet, then the condition will be $L = \lambda/4$ with half the frequency.) Hence $\omega = 2\pi(v/L)$ where $v = (F/\rho)^{1/2}$ for the stretched string, where F is the tension and ρ the mass per unit length.

The speed of sound in a solid material is $v = (Y/\rho)^{1/2}$, with Y the Young's modulus. Young's modulus represents force per unit area (pressure stress) per fractional deformation (strain). Young's modulus is therefore a fundamental rigidity parameter of a solid, related to the bonding of its atoms. For brass, $Y = 90\,GPa = 90 \times 10^9\,N/m^2$. (This means, e.g., that a pressure F/A of 101 kPa applied to one end of a brass bar of length $L = 0.1$ m would compress its length by $\Delta l = LF/YA = 11\,\mu m$.) Note that $Y = 90\,GPa$ and $\rho = 10^4\,kg/m^3$, values similar to brass, correspond to a speed of sound $v = 3000$ m/s. On this basis the longitudinal resonant frequency of a 0.1 m brass rod is $f = v/2L = 15$ kHz. This frequency is in the ultrasonic range.

If one could shorten a brass rod to 0.1 micron in length, the corresponding frequency would be 15 GHz, which corresponds to an electromagnetic wave with 2 cm wavelength. This huge change in frequency will allow completely different applications to be addressed, achieved simply by changing the size of the device!

A connection between macroscopic and nanometer scale descriptions can be made by considering a linear chain of N masses m spaced by springs of constants K, of length a. The total length of the linear chain is thus $L = Na$.

Vibrations on a Linear Atomic Chain of length L= Na

On a chain of N masses of length L, and connected by springs of constant K, denote the longitudinal displacement of the nth mass from its equilibrium position by u_n. The differential equation (Newton's Second Law) $F = ma$ for the nth mass is

$$md^2u_n/dt^2 + K(u_{n+1}-2u_n-u_{n-1}) = 0. \tag{2.4}$$

A traveling wave solution to this equation is

$$u_n = u_o\cos(\omega t + kna). \tag{2.5}$$

Here kna denotes $kx = 2\pi x/\lambda$, where k is referred to as the wave number. Substitution of this solution into the difference equation reveals the auxiliary condition $m\omega^2 = 4K\sin^2(ka/2)$. This "dispersion relation" is the central result for this problem. One sees that the allowed frequencies depend upon the wavenumber $k = 2\pi/\lambda$, as

$$\omega = 2(K/m)^{1/2}\left|\sin(ka/2)\right|. \tag{2.6}$$

The highest frequency, $2(K/m)^{1/2}$ occurs for $ka/2 = \pi/2$ or $k = \pi/a$; where the wavelength $\lambda = 2a$, and nearest neighbors move in opposite directions. The smallest frequency is at $k = \pi/Na = \pi/L$, which corresponds to $L = \lambda/2$. Here one can use the expansion of $\sin(x) \cong x$ for small x. This gives $\omega = 2(K/m)^{1/2}ka/2 = a(K/m)^{1/2}k$, representing a wave velocity $\omega/k = v = a(K/m)^{1/2}$.

Comparing this speed with $v = (Y/\rho)^{1/2}$ for a thin rod of Young's modulus Y, and mass density ρ, we deduce that $Y/\rho = Ka^2/m$. Here K and a are, respectively, the spring constant, and the spacing a of the masses. [1]

Young's modulus can thus be expressed in microscopic quantities as $Y = \rho Ka^2/m$ if the atoms have spacing a, mass m, and the interactions can be described by a spring constant K.

Table 2.1 Mechanical properties of some strong solids

Material	Young's modulus Y (GPa)	Strength (GPa)	Melting point (K)	Density ρ (kg/m^3)
Diamond	1050	50	1800	3500
Graphite	686	20	3300	2200
SiC	700	21	2570	3200
Si	182	7	1720	2300
Boron	440	13	2570	2300
Al$_2$O$_3$	532	15	2345	4000
Si$_3$N$_4$	385	14	2200	3100
Tungsten	350	4	3660	19300

A cantilever of length L clamped at one end and free at the other, such as a diving board, resists transverse displacement y (at its free end, $x = L$) with a force $-Ky$. The effective spring constant K for the cantilever is of interest to designers of scanning tunneling microscopes and atomic force microscopes, as well as to divers. The resonant frequency of the cantilever varies as L^{-2} according to the relation $\omega = 2\pi (0.56/L^2)(YI/\rho A)^{1/2}$. Here ρ is the mass density, A the cross section area and I is the moment of the area in the direction of the bending motion. If t is the thickness of the cantilever in the y direction, then $I_A = \int A(y)y^2 dy = wt^3/12$ where w is the width of the cantilever. It can be shown that $K = 3YI/L^3$. For cantilevers used in scanning tunneling microscopes the resonant frequencies are typically $10\,\text{kHz} - 200\,\text{kHz}$ and the force constant K is in the range $0.01 - 100$ Newtons/m. It is possible to detect forces of a small fraction of a nanoNewton (nN).

Cantilevers can be fabricated from Silicon using photolithographic methods. Shown in Figure 1.1 is a "nanoharp" having silicon "wires" of thickness $50\,\text{nm}$ and lengths L varying from $1000\,\text{nm}$ to $8000\,\text{nm}$. The silicon rods are not under tension, as they would be in a musical harp, but function as doubly clamped beams. The resonant frequency for a doubly clamped beam differs from that of the cantilever, but has the same characteristic L^{-2} dependence upon length: $\omega = (4.73/L)^2(YI/\rho A)^{1/2}$. The measured resonant frequencies in the nanoharp structure range from $15\,\text{MHz}$ to $380\,\text{MHz}$.

The largest possible vibration frequencies are those of molecules, for example, the fundamental vibration frequency of the CO molecule is $6.42 \times 10^{13}\,\text{Hz}$ ($64.2\,\text{THz}$). Analyzing this vibration as two masses connected by a spring, the effective spring constant is $1860\,\text{N/m}$.

2.2
Scaling Relations Illustrated by a Simple Harmonic Oscillator

Consider a simple harmonic oscillator (SHO) such as a mass on a spring, as described above, and imagine shrinking the system in three dimensions. As stated, $m \alpha L^3$ and $K \alpha L$, so $\omega = (K/m)^{1/2} \alpha L^{-1}$.

It is easy to see that the spring constant scales as L if the "spring" is taken a (massless) rod of cross section A and length L described by Young's modulus $Y = (F/A)/(\Delta L/L)$. Under a compressive force F, $\Delta L = -(LY/A)F$, so that the spring constant $K = LY/A$, αL.

A more detailed analysis of the familiar coiled spring gives a spring constant $K = (\pi/32R^2)\mu_S d^4/\ell$, where R and d, respectively, are the radii of the coil and of the wire, μ_S is the shear modulus and ℓ the total length of the wire. [1] This spring constant K scales in three dimensions as L^1.

Insight into the typical scaling of other kinetic parameters such as velocity, acceleration, energy density, and power density can be understood by further consideration of a SHO, in operation, as it is scaled to smaller size. A reasonable quantity to hold constant under scaling is the strain, x_{max}/L, where x_{max} is the amplitude of the motion and L is length of the spring. So the peak velocity of the mass $v_{max} = \omega x_{max}$

which is then constant under scaling: $v\alpha L^0$, since $\omega \alpha L^{-1}$. Similarly, the maximum acceleration is $a_{max} = \omega^2 x_{max}$, which then scales as $a\alpha L^{-1}$. (The same conclusion can be reached by thinking of a mass in circular motion. The centripetal acceleration is $a = v^2/r$, where r is the radius of the circular motion of constant speed v.) Thus for the oscillator under isotropic scaling the total energy $U = 1/2\, Kx_{max}^2$ scales as L^3.

In simple harmonic motion, the energy resides entirely in the spring when $x = x_{max}$, but has completely turned into kinetic energy at $x = 0$, a time $T/4$ later. The spring then has done work U in a time $1/\omega$, so the power $P = dU/dt$ produced by the spring is $\alpha\, \omega U$, which thus scales as L^2. Finally, the power per unit volume (power density) scales as L^{-1}. The power density strongly increases at small sizes. These conclusions are generally valid as scaling relations.

2.3
Scaling Relations Illustrated by Simple Circuit Elements

A parallel plate capacitor of area A and spacing t gives $C = \varepsilon_o A/t$, which under isotropic scaling varies as L. The electric field in a charged capacitor is $E = \sigma/\varepsilon_o$, where σ is the charge density. This quantity is taken as constant under scaling, so E is also constant. The energy stored in the charged capacitor $U = Q^2/2C = (1/2)\, \varepsilon_o E^2 At$, where At is the volume of the capacitor. Thus U scales as L^3. If a capacitor is discharged through a resistor R, the time constant is $\tau - RC$. Since the resistance $R = \rho\ell/A$, where ρ is the resistivity, ℓ the length, and A the constant cross section of the device, we see that R scales as L^{-1}. Thus the resistive time constant RC is constant (scales as L^0). The resistive electrical power produced in the discharge is $dU/dt = U/RC$, and thus scales as L^3. The corresponding resistive power density is therefore constant under scale changes.

Consider a long wire of cross section A carrying a current I. Ampere's Law gives $B = \mu_o I/2\pi R$ as the (encircling) magnetic field B at a radius R from the wire. Consider scaling this system isotropically. If we express $I = AE/\rho$, where E is the electric field in the wire, assumed constant in the scaling, and ρ is the resistivity, then B scales as L. The assumption of a scale-independent current density driven by a scale-independent electric field implies that current I scales as L^2. The energy density represented by the magnetic field is $\mu_o B^2/2$. Therefore the magnetic energy U scales as L^5. The time constant for discharge of a current from an inductor L' through a resistor R is L'/R. The inductance L' of a long solenoid is $L' = \mu_o n^2 A\ell$, where n is the number of turns per unit length, A is the cross section and ℓ the length. Thus inductance L' scales as length L, and the inductive time constant L'/R thus scales as L^2.

For an LC circuit the charge on the capacitor $Q = Q(0)\cos[(C/L)^{1/2}t]$. The radian resonant frequency $\omega_{LC} = (C/L)^{1/2}$ thus scales as L^0.

2.4
Thermal Time Constants and Temperature Differences Decrease

Consider a body of heat capacity C (per unit volume) at temperature T connected to a large mass of temperature $T = 0$ by a thermal link of cross section A, length L and thermal conductivity k_T. The heat energy flow dQ/dt is $k_T AT/L$ and equals the loss rate of thermal energy from the warm mass, $dQ/dt = CV dT/dt$. The resulting equation $dT/T = -(k_T A/LCV)dt$ leads to a solution $T = T(0)\exp(-t/\tau_{th})$, where $\tau_{th} = LCV/k_T A$. Under isotropic scaling τ_{th} varies as $L^2 C/k_T$. Thermal time constants thus strongly decrease as the size is reduced. As an example, a thermal time constant of a few μs is stated [2] for a tip heater incorporated into a 200 kHz frequency AFM cantilever designed for the IBM "Millipede" 1024 tip AFM high density thermomechanical memory device. The heater located just above the tip, mounted at the vertex of two cantilever legs each having dimensions $50 \times 10 \times 0.5$ micrometers.

In steady state with heat flow dQ/dt, we see that the temperature difference T is $T = (dQ/dt)(L/k_T A)$. Since the mechanical power dQ/dt scales as L^2, this result implies that the typical temperature difference T scales, in three dimensions, as L. Temperature differences are reduced as the size scale is reduced.

2.5
Viscous Forces Become Dominant for Small Particles in Fluid Media

The motion of a mass in a fluid, such as air or water, eventually changes from inertial to diffusive as the mass of the moving object is reduced. Newton's Laws (inertial) are a good starting point for the motions of artillery shells and baseballs, even though these masses move through a viscous medium, the atmosphere. The first corrections for air resistance are usually velocity-dependent drag forces. A completely different approach has to taken for the motion of a falling leaf or for the motion of a microscopic mass in air or in water.

The most relevant property of the medium is the viscosity η, defined in terms of the force $F = \eta v A/z$ necessary to move a flat surface of area A parallel to an extended surface at a spacing z and relative velocity v in the medium in question. The unit of viscosity η is the Pascal-second (one Pascal is a pressure of 1 N/m^2). The viscosity of air is about 0.018×10^{-3} Pa·s, while the value for water is about 1.8×10^{-3} Pa·s. The traditional unit of viscosity, the Poise, is 0.1 Pa·s in magnitude.

The force needed to move a sphere of radius R at a velocity v through a viscous medium is given by Stokes' Law,

$$F = 6\pi\eta R v. \tag{2.7}$$

This is valid only for very small particles and small velocities, under conditions of streamline flow such that the Reynolds number $N_{Reynolds}$ is less than approximately 2000. $N_{Reynolds}$, which is dimensionless, is defined as $N_{Reynolds} = 2R\rho v/\eta$, where R is the radius, ρ the mass density, v the velocity and η the viscosity.

The fall, under the acceleration of gravity g, of a tiny particle of mass m in this regime is described, following Stokes' Law, by a limiting velocity obtained by setting F (from equation 2.6) equal to mg. This gives

$$v = mg/6\pi\eta R. \tag{2.8}$$

As an example, a particle of $10\,\mu m$ radius and density $2000\,kg/m^3$ falls in air at about $23\,mm/s$, while a $15\,nm$ particle of density $500\,kg/m^3$ will fall in air at about $13\,nm/s$. In the latter case one would expect random jostling forces $f(t)$ on the particle by impacts with individual air molecules (Brownian motion) to be present as well as the slow average motion. Newton's laws of motion as applied to the motion of artillery shells are not useful in such cases, nor for any cases of cells or bacteria in fluid media.

An appropriate modification of Newton's Second Law for such cases is the Langevin equation [3]

$$F_{ext} + f(t) = [4\pi\rho R^3/3]d^2x/dt^2 + 6\pi\eta R\, dx/dt. \tag{2.9}$$

This equation gives a motion $x(t)$ which is a superposition of drift at the terminal velocity (resulting, as above, from the first and last terms in the equation) and the stochastic diffusive (Brownian) motion represented by $f(t)$.

In the absence of the external force, the diffusive motion can be described by

$$P(x,t) = (4\pi Dt)^{-3/2}\exp(-x^2/4Dt), \tag{2.10}$$

where

$$D = kT/6\pi\eta R \tag{2.11}$$

is the diffusivity of the particle of radius R in a fluid of viscosity η at temperature T. Consideration of the exponential term allows one to define the "diffusion length" as

$$x_{rms} = (4Dt)^{1/2}. \tag{2.12}$$

Returning to the fall of the $15\,nm$ particle in air, which exhibits a drift motion of $13\,nm$ in one second, the corresponding diffusion length for $300\,K$ is $x_{rms} = 2D^{1/2} = 56\,\mu m$. It is seen that the diffusive motion is dominant in this example.

The methods described here apply to slow motions of small objects where the related motion of the viscous medium is smooth and not turbulent. The analysis of diffusion is more broadly applicable, for example, to the motion of electrons in a conductor, to the spreading of chemical dopants into the surface of a silicon crystal at elevated temperatures, and to the motion of perfume molecules through still air.

In the broader but related topic of flying in air, a qualitative transition in behavior is observed in the vicinity of $1\,mm$ wingspan. Lift forces from smooth flow over airfoil surfaces, which derive from Bernoulli's principle, become small as the scale is reduced. The flight of the bumblebee is not aerodynamically possible, we are told, and the same conclusion applies to smaller flying insects such as mosquitoes and gnats. In these cases the action of the wing is more like the action of an oar as it is forced against the relatively immovable water. The reaction force against moving the viscous and massive medium is the force that moves the rowboat and also the force that lifts the bumblebee.

No tiny airplane can glide, another consequence of classical scaling. A tiny airplane will simply fall, reaching a terminal velocity that becomes smaller as its size is reduced.

These considerations only apply when one is dealing with small particles in a liquid or a gas. They do not apply in the prospect of making smaller electronic devices in silicon, for example.

2.6
Frictional Forces can Disappear in Symmetric Molecular Scale Systems

Viscous and frictional forces are essentially zero in a nanotechnology envisioned by Drexler [4] in which moving elements such as bearings and gears of high symmetry are fashioned from diamond-like "diamondoid" covalently bonded materials. (To be sure, these are only computer models, no such structures have been fabricated.) The envisioned nanometer scaled wheels and axles are precisely self-aligned in vacuum by balances of attractive and repulsive forces, with no space for any fluid. Such moving parts basically encounter frictional forces only at a much lower level than fluid viscous forces.

There are natural examples of high symmetry nested systems. One example is provided by nested carbon nanotubes, which are shown in Figures 2.1 and 2.2 (after Cumings and Zettl, [5]). Nanotubes are essentially rolled sheets of graphite, which

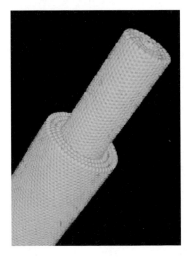

Figure 2.1 Nested carbon nanotubes [5]. This is a computer generated image. Zettl [5] has experimentally demonstrated relative rotation and translation of nested nanotubes, a situation very similar to that in this image. The carbon–carbon bonding is similar to that in graphite. The spacer between the tubes is simply vacuum

have no dangling bonds perpendicular to their surfaces. Graphite is well known for its lubricating properties, which arise from the easy translation of one sheet against the next sheet. It is clear that there are no molecules at all between the layers of graphite, and the same is true of the nanotubes. The whole structure is simply made of carbon atoms. The medium between the very closely spaced moving elements is vacuum.

A second example in nature of friction-free motion may be provided by the molecular bearings in biological rotary motors. Such motors, for example, rotate flagella (propellers) to move cells in liquid. The flagellum is attached to a shaft which rotates freely in a molecular bearing structure. The rotation transmits torque and power and the motors operate continuously over the life of the cell, which suggests a friction-free molecular bearing. This topic will be taken up again in Chapter 3.

The double nanotube structure is maintained in its concentric relation by forces between the carbon atoms in the inner and outer tubes. These forces are not easy to fully characterize, but one can say that there is negligible covalent bonding between individual atoms on the adjacent tubes. Presumably there are repulsive overlap forces between atoms on adjacent tubes, such that a minimum energy (stable configuration) occurs when the tubes are parallel and coaxial. Attractive forces are presumably of the van der Waals type, and again the symmetry would likely favor the concentric arrangement. In the elegant experiments of Zettl [5], it was found that a configuration as shown in Figure 2.2 would quickly revert to a fully nested configuration when the displaced tube was released. This indicates a net negative energy of interaction between the two tubes, which will pull the inner tube back to full nesting. This is in accordance with the fact that the attractive van der Waals force is of longer range than the repulsive overlap force, which would be expected to have a negative exponential dependence on the spacing of the two tubes.

The structures of Figures 2.1 and 2.2 make clear that there will be a corrugated potential energy function with respect to relative translation and relative rotation, with periodicity originating in the finite size of the carbon atoms. The barriers to rotation and translation must be small compared to the thermal energy kT, for free rotation and translation to occur, and also for apparently friction-free motions to occur without damage to the structures. Incommensurability of the two structures will reduce such locking tendencies. An analysis of incommensurability in the

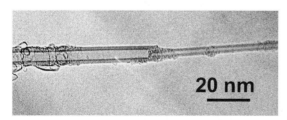

Figure 2.2 TEM image of partially nested nanotubes, after relative translation (Cumings and Zettl [6]). It was found that the inner tube could repeatedly be slid and rotated within the outer tube, with no evidence of wear or friction. Attractive forces very rapidly pulled a freed tube back into its original full nesting

design of molecular bearings has been given by Merkel [7]. Again, there are no known methods by which any such structures can be fabricated, nor are there immediate applications.

Rotational and translational relative motions of nested carbon nanotubes, essentially free of any friction, are perhaps prototypes for the motions envisioned in the projected diamondoid nanotechnology. The main question is whether elements of such a nanotechnology can ever be fabricated in an error-free fashion so that the unhindered free motions can occur.

References

[1] For more details on this topic, see A. Guinier and R. Jullien, *The Solid State*, (Oxford, New York, 1989).

[2] P. Vettiger, M. Despont, U. Drechsler, U. Durig, W. Haberle, M.I. Lutwyche, H.E. Rothuizen, R. Stutz, R. Widmer, and G.K. Binnig, IBM J. Res. Develop. **44**, 323 (2000).

[3] E. A. Rietman, *Molecular Engineering of Nanosystems*, (Springer, New York, 2001), p. 55.

[4] K. Eric Drexler, *Nanosystems*, (Wiley, New York, 1992).

[5] Image courtesy of Zettl Research Group, University of California at Berkeley and Lawrence Berkeley National Laboratory.

[6] Reprinted with permission from J. Cummings and A. Zettl, Science **289**, 602–604 (2000). Copyright 2000 AAAS.

[7] R. C. Merkle, Nanotechnology **8**, 149 (1997).

3
What are Limits to Smallness?

3.1
Particle (Quantum) Nature of Matter: Photons, Electrons, Atoms, Molecules

The granular nature of matter is the fundamental limit to making anything arbitrarily small. No transistor smaller than an atom, about 0.1 nm, is possible. That chemical matter is composed of atoms is well known! In practice, of course, there are all sorts of limits on assembling small things to an engineering specification. At present there is hardly any systematic approach to making arbitrarily designed devices or machines whose parts are much smaller than a millimeter! A notable exception is the photolithographic technology of the semiconductor electronics industry which make very complex electronic circuits with internal elements on a much smaller scale, down to about 100 nm. However, this approach is essentially limited to forming two-dimensional planar structures.

It is not hard to manufacture Avogadro's number of H_2O molecules, which are individually less than one nm in size. One can react appropriate masses of hydrogen and oxygen, and the H_2O molecules will "self-assemble" (but stand back!). But to assemble even 1000 of those H_2O molecules (below 0 °C) in the form, e.g., of the letters "IBM", is presently impossible. Perhaps this will not always be so.

The most surprising early recognition of the granularity of nature was forced by the discovery that light is composed of particles, called photons, whose precise energy is $h\nu$. Here h is Planck's constant, 6.6×10^{-34} J·s, and ν is the light frequency in Hz. The value of the fundamental constant h was established by quantitative fits to the measurements of the classically anomalous wavelength distribution of light intensity emitted by a body in equilibrium at a temperature T, the so-called "black body spectrum" [1].

The energy of a particle of light in terms of its wavelength, λ, is

$$E = h\nu = hc/\lambda. \tag{3.1}$$

A convenient approach to calculating E, in eV, giving λ in nm, involves remembering that the product $hc = 1240$ eV·nm.

Nanophysics and Nanotechnology: An Introduction to Modern Concepts in Nanoscience. Edward L. Wolf
Copyright © 2004 WILEY-VCH Verlag GmbH & Co. KGaA, Weinheim
ISBN: 3-527-40407-4

Later, it was found that electrons are emitted in a particular discrete manner from a metal surface illuminated by light of wavelength $\lambda = c/\nu$. The maximum kinetic energy of the photoelectrons, K, was measured to be

$$K = (hc/\lambda) - \phi, \tag{3.2}$$

where ϕ, the work function, is characteristic of each metal but usually in the range of a few electron volts (eV) [2]. The important thing was that the same value of h was obtained from this experiment as from the light spectrum analysis of Planck. It was found that the energy of the electrons (as distinct from their number) was not affected by the intensity of the light but only by its frequency. So it was clear that the light was coming in discrete chunks of energy, which were completely absorbed in releasing the electron from the metal (the binding energy of the electron to the metal is ϕ).

As far as electrical charge Q is concerned, the limit of smallness is the charge $-e$ of the single electron, where $e = 1.6 \times 10^{-19}$ C. This is an exceedingly small value, so the granularity of electrical charge was not easily observed. For most purposes electrical charge can be considered to be a continuously distributed quantity, described by volume density ρ or surface density σ.

The first measurement of the electron charge e was made by the American physicist Robert Millikan, who carefully observed [3] the fall of electrically charged microscopic droplets of oil in air under the influence of gravity and a static electric field. Following the application of Stokes' Law, (see equations 2.7 and 2.8) the velocity of fall in air is given by

$$v = (mg + neE)/(6\pi\nu R), \tag{3.3}$$

where n is the number of electron charges on the droplet. By making measurements of the electric field E needed to make the velocity v of a single drop come to zero, as its charge number n changed in his apparatus, Millikan was able to deduce the value of the electron charge e.

These historical developments, seminal and still highly relevant to the origins of nanophysics, are summarized in introductory sections of [6] and [7] which are quickly available and inexpensive.

3.2
Biological Examples of Nanomotors and Nanodevices

Biology provides examples of nanometer scale motors and electrical devices, which can be seen as limits of smallness. If nature can make these (only recently perceived) nanoscale machines, why, some ask, cannot human technology meet and eventually exceed these results? It is certainly a challenge.

The contraction of muscle occurs through the concerted action of large numbers of muscle myosin molecules, which "walk" along actin filaments in animal tissue.

Other sorts of motors rotate flagella, providing motion of bacteria through liquid media. Myosin, Kinesin, and rotary motors for flagella appear in extremely primitive life forms. It is believed that original forms of these proteins were present in single cell organisms appearing about a billion years ago. Simpler motors (similarly ancient) resembling springs are exemplified by the spasmoneme in *Vorticella*.

Electrically controlled valves in biology are exemplified by ion channels. One of the most studied is the voltage-gated potassium channel. This protein assembly controllably allows potassium ions to cross the lipid membrane of a neuron, generating nerve impulses.

These ion channels can be compared to transistors, in that a voltage controls a current flow. Ion channels, embedded in lipid cell walls, are truly nanometer scale electrically controlled gate devices.

3.2.1
Linear Spring Motors

An example of a biological motor, the spasmoneme spring, is shown in Figure 3.1 [6,7].

This system, first observed by the famous microscopist Leeuwenhoek in 1676, has a large literature. When extended, the spring may be mm in length. When exposed to calcium, which neutralizes the net negative charge in the extended state, the stalk contracts in a few ms to 40% of its length, at velocities approaching 8 cm/s.

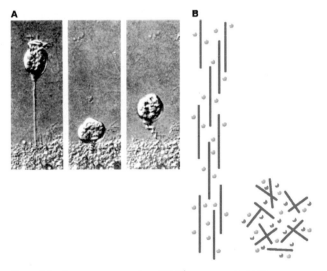

Figure 3.1 Spasmoneme spring. (A). The spasmoneme in *Vorticella* in its fully extended (left), fully contracted (middle), and partly extended (right) states [6]. (B). The extended spring state (left) consists of aligned filaments held apart by negative charges (dots). On the right, with plus and minus charges (dots) equally likely, the stalk collapses to a rubber-like state

In consideration of the viscous drag forces that are involved, as described in Chapter 2, noting that the diameter of the head is as much as 50 μm, the force of contraction is estimated to be on the order of 10 nN, and the power is of order 100 pW.

As sketched in Figure 3.1 (B), the stalk is modeled [6] to consist of a bundle of filaments, each about 2 nm in diameter, which are roughly aligned but only weakly cross-linked. The extended structure is net negatively charged, forced into this extended state simply by Coulombic repulsion within the linear constraint of the filament bundle. Many further subtleties of this situation are known, including a helical aspect, so that the spasmoneme rotates as it contracts.

A somewhat similar spring motor is believed to operate in sperm cells exemplified by the horseshoe crab *Limulus polyphemus* in which a 60 μm finger, called the "acrosomal process", extends quickly across a jelly-like barrier to accomplish fertilization of egg plasma. Again, the extension motion seems clearly to be the result of Coulombic repulsion, as the ionic charge in the system is changed. There seems to be an opportunity to reanalyze some of this careful and extensive literature in more simple electrostatic terms.

The microbiological literature scarcely contains mention of electrostatic forces, yet these must be the essential origin of motions in bio-molecular motors, from a perhaps naïve, but unavoidably basic point of view. (Magnetic forces are certainly to be excluded, in the context of motors, but not in the context of sensors, as we will see below.)

Interdisciplinary approaches involving applied physicists familiar with semiconductor charge layers, screened electrostatic forces, etc., working with biologists, are likely to be very productive.

3.2.2
Linear Engines on Tracks

The spring and ratchet systems, structurally ill-defined but clearly electrostatic in mechanism, contrast with well defined linear and rotary engines, which move in stepwise fashions, however, with mechanisms that remain unclear. In this discussion we mention the linear motors myosin and kinesin [8]; and an example of a rotary motor, suitable for driving a flagellum, in this case F_1-adenosine triphosphate synthase. In contrast to the spring systems, the energetics of these linear and rotary motors is definitely tied to the conversion (hydrolysis) of ATP to ADP. The efficiency of conversion of chemical energy to mechanical energy appears to be close to unity.

Footnote 33 to [8] states "The work efficiencies of kinesin and myosin (each 50–60% efficient) are much greater than that of an automobile (10–15%)".

There is no quarrel with this statement, but it may not be appreciated that the biological and internal combustion engines are of entirely different types. The automobile engine is a heat engine, as described by the laws of thermodynamics, with efficiency that depends essentially on the ratio of the input and exhaust gas temperatures on the Kelvin scale, where room temperature is about 300 K. The *compression ratio* is a quantity to be maximized, for example, for an efficient diesel or gasoline internal combustion engine, and the mathematics for the ideal or limiting case is based on the Carnot cycle. The compression ratio has nothing to do with biological engines.

The biological motors, in contrast, provide direct conversion of chemical to mechanical energy. Temperature plays no essential role, and the system is more like a chemical battery except that the output is realized in mechanical rather than electrical terms.

The linear engines, myosin and kinesin, move along protein polymers, (actin and microtubules, respectively), in quantized steps, using up a fixed number of ATP molecules per step [8]. The actin filaments, for example, are about 5nm in diameter and 1 – 4 micrometers in length. The essence of the step-wise motion seems to be a change in *conformation* (shape or configuration) of the large motor molecule, energetically enabled by the hydrolysis of ATP, which allows the motor to move one well-defined step along its substrate. The rotary motor is less transparent in its operation. Yet it seems likely that these cases may eventually all be understood in essentially electrostatic terms.

A summary of various types of biological cellular engines is provided in Table 3.1 [6].

Table 3.1 Cellular engines of biology [6]. The performance of various cellular engines is compared with thermal energy (kT) and a typical automobile engine. Calculations for the specific power are based on the molecular weight of the smallest unit of the engine. Thus, molecular motors and polymerization-based engines are more powerful than the cellular structures in which they are found; for example, compare myosin to striated muscle

Engine	Velocity ($\mu m\ s^{-1}$)	Force (dynes)	Specific Power (erg s^{-1} g^{-1})
kT (thermal energy)	–	4×10^{-7} dyne nm	–
Actin polymerization [9]	1.0	1×10^{-6}	1×10^9
Microtubule polymerization [10]	0.02	4×10^{-7}	5×10^8
Myosin II [11]	4.0	1×10^{-6}	2×10^8
Kinesin [12]	1.0	6×10^{-7}	7×10^7
Vorticellid spasmoneme [13]	8×10^4	1×10^{-3}	4×10^7
Automobile engine, typical [14]			3×10^6
Striated muscle [14]			2×10^6
Bacterial flagellar motor [15,16]	100 Hz	4.5×10^{-11} dyne cm	1×10^6
Thyone acrosomal reaction [17]	6–9	5×10^{-4}	1×10^5
Limulus acrosome reaction	10	1×10^{-6}	1×10^4
Eukaryotic flagellum [14]	–		3×10^2
Mitotic spindle [14]	2.0	1×10^{-5}	

Models for the movement of muscle myosin and kinesin are given in Figure 3.2 [8]. In both cases the motor moves along a strong filament which is part of the internal structure of the cell. The function of kinesin is to move material within the cell, the upward extended filament in Figure 3.2 B is attached to such a load.

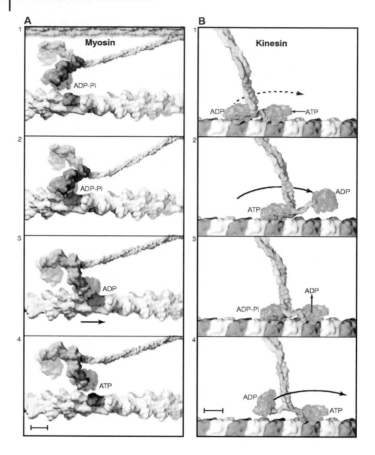

Figure 3.2 Models of the motion of muscle myosin and conventional kinesin [8]. (A). Frame 1: Muscle myosin is a dimer of two identical motor heads, anchored to a thick filament (top). Frames 2–4 show docking of heads on actin filament (lower), which serves to move the actin. The motion is about 10 nm per ATP hydrolyzed. (B). The two heads of the kinesin dimer move along a tubulin filament as indicated in Frames 1–4. In these frames the coiled coil extends to the top and to the attached cargo. The step length is about 8 nm. The head motion is associated with hydrolysis of ATP to ADP. The scale bar in A is 6 nm; that in B is 4 nm. (Taken from [8])

The details of the muscle myosin movement have been recently confirmed [18,19]. It is now definitely known that the trailing muscle myosin head leaps ahead of the forward fixed head, in a motion similar to bipedal walking, or hand-over-hand rope climbing.

3.2.3
Rotary Motors

Two examples of biomolecular rotary motors are F_1-adenosine triphosphate synthase (F_1-ATPase) and ATP Synthase (F_0F_1) [20,21]. These motors depicted in Figures 3.3 and 3.4, are similar to motors that rotate flagella in bacteria and appear to have originated in cells about a billion years ago. They have evolved and serve many different purposes. These motors occur mounted in the wall of a cell and transmit rotational torque and power across the cell wall. The cell wall is a lipid hydrophobic membrane impervious to flows of liquids and ions and generally sealing the interior of the cell from its environment. The cell is internally structured with rigid actin filaments and tubulins, which act as tracks for the linear engines mentioned above.

The F_1-adenosine triphosphate synthase (F_1-ATPase) motor is shown in Figure 3.3 [20]. This rotary engine, which is found in mitochondria and bacterial membranes, "couples with an electrochemical proton gradient and also reversibly hydrolyzes ATP to form the gradient" [20]. It is shown in these experiments that this motor does work by rotating an attached actin filament through a viscous fluid. The work done by this motor in some biological contexts may be understood as $W = e\Delta V$, where e is the charge of the proton (hydrogen ion H^+) and ΔV is electrostatic potential difference. In such a case we have a "proton pump".

ATP Synthase (F_0F_1) motors from *e. coli* were attached to a glass coverslip, using the His tag linked to the lower end of the α subunit (lower portion of Figure 3.3). In

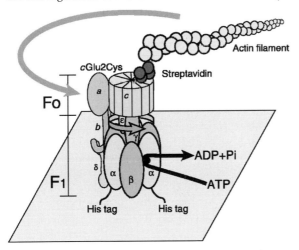

Figure 3.3 Fluorescently labeled actin filament permits observation of rotation of the *c* subunit in ATP Synthase (F_0F_1) [20]. A *c* subunit of Glu^2 was replaced by cysteine and then biotinylated to bind streptavidin and the actin filament. The γ, ε, and *c* units are thus shown to be a rotor, while the α, β, δ, *a*, and *b* complex is the stator. The rotation rate of the actin filament in the viscous medium was found to depend upon its length. Rotational rates in the range 0.5 Hz – 10 Hz were measured, consistent with a torque τ of 40 pN·nm

this motor, the γ, ε, and c units rotate together, as shown in this work. This was demonstrated by attaching the actin filament to the top of the motor, the c unit. Attachment was accomplished in an elegant fashion making use of Streptavidin as shown. The actin filament was tagged with molecules which give characteristic fluorescence when illuminated, in the fashion described above in connection with quantum dots. Video microscope pictures of the fluorescing actin filament revealed counter clockwise rotation (see arrows), when the ambient solution contained 5 mM Mg ATP. The rotation rate of the actin filament in the viscous medium was found to depend upon its length. Rotational rates in the range 0.2 Hz – 3.5 Hz were measured, consistent with a torque τ of 40 pN·nm.

Analysis of the rotational torque exerted in the viscous fluid was based on the expression

$$\tau = (4\pi/3)\omega\eta L^{3}[\ln(L/2r) - 0.447]^{-1}. \tag{3.4}$$

Here ω is the rotation rate, η is the viscosity 10^{-3} Pa·s; L and r, respectively, are the length and radius of the actin filament. The actin filament radius r is taken as 5 nm. The measurements indicate a constant torque τ of about 40 pN·nm.

These experiments demonstrate an accumulated virtuosity in integrating biological, chemical, and microphysical insight, experimental design, and measurements. To form the experimental structure, alone, is a great accomplishment. The "fluorescently labeled actin filament-biotin-streptavidin complex" by which the rotation was observed in [20] is an excellent example.

The durability of such motors when harnessed as propeller-turners has been demonstrated by Soong et al. [21]. It is common to use biological products in human endeavors, but certainly very uncommon to detach a molecular motor to perform work in the inanimate world.

Figure 3.4 [21] shows a biomolecular rotary motor, F_1-adenosine triphosphate synthase (F_1-ATPase), harnessed to turn a nanopropeller, allowing measurement of the power output and efficiency of the motor. The motor itself is sketched in panel B of Figure 3.4, and its deployment to turn the propeller is sketched in panel D. The assembled motors are immersed in a buffer solution containing either 2 mM Na_2ATP (which fuels rotation), or 10 mM NaN_3 (Sodium azide) which stops the motion. Panel A shows a portion of an array of nickel posts, typically 50 to 120 nm in diameter and about 200 nm high, spaced by about 2.5 μm, as deposited on a cover glass. The inset to A shows an 80 nm diameter post in a particular experiment.

The motors were mounted on the posts in order to space the propellers away from the base plane. This allows calculations of the work required to turn the propeller in the viscous fluid to be simplified.

The Ni propellers were fabricated, coated suitably to bind to the top of the rotary engines, and suspended in a buffer solution which flowed past the array of engines mounted on Ni posts. In the reported experiment [21] about 400 propellers led to observation of five that rotated continuously in a counter clockwise direction. The attachment points along the propeller length of the rotor axis showed a random distribution, but the lengths L_1 and L_2 of the two sides from the rotor axis were mea-

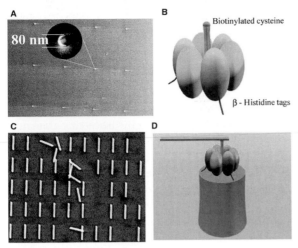

Figure 3.4 Bio-molecular rotary motor powered propellers. [21]
(A) 80 nm Ni post from array. (B) Schematic view of F_1-ATPase
molecular motor. (C) Array of Ni propellers, 750 nm – 1400 nm
in length, 150 nm in diameter. (D) Schematic view of one
assembled device from array. Rotation in fluid of propeller
(0.8 – 8.3 rps) fueled with ATP is 50% efficient

sured. The observed speeds of rotation depended upon the values of L_1 and L_2, ranging from 0.8 Hz to 8.3 Hz. It was found that some of the propellers ran for almost 2.5 hours while ATP was present in the cell. The rotation could be stopped by adding sodium azide to the cell.

An analysis of the work done by the rotating propellers was based on an expression for the force per unit length dF/dR exerted by the viscous medium on an element dR of propeller at radius R from the axis, on the assumption that the height h above a flat surface is small:

$$dF/dR = 4\pi\eta\omega R[\cosh^{-1}(h/r)]^{-1}. \tag{3.5}$$

In this expression η is the viscosity of the medium, 10^{-3} Pa·s, ω is the angular velocity of rotation, R the radius from the axis of rotation, h the pillar height (200 nm), and r half the width of the propeller (75 nm). Integration of this expression on R led to the expression for the torque τ

$$\tau = 4\pi\eta\omega(L_1^3 + L_2^3)\,[3\cosh^{-1}(h/r)]^{-1}. \tag{3.6}$$

Here L_1 and L_2 are the lengths of the propeller sections extending from the rotational axis.

The determined values of the torque τ were 20 pN·nm for the 750 nm propellers (8 Hz, attached 200 nm from the end) and 19 pN·nm for the 1400 nm propellers (1.1 Hz, attached 350 nm from the end). The energy for one turn of these propellers is then 119 – 125 pN·nm. The energy released by hydrolysis of three ATP molecules is reported as about 240 pN·nm, leading to an efficiency of these motors of about 50% [21].

The experiments depicted in Figures 3.3 and 3.4 prove that engines of biology can act as stand-alone machines which function in suitably buffered environments, living or dead, if given ATP. The question: "How many pounds of muscle myosin would be needed to power a typical SUV?", if irreverent, is not entirely irrelevant. After all, large mammals, including whales and elephants, move by muscle myosin. The basic question, (solved by nature), is how to organize these tiny elemental engines to pull together.

It has been suggested that these tiny motors (available, and abundantly tested, proven over millions of years) might fill a gap in the development of nanoelectromechanical systems (NEMS), namely, for suitable power sources. As is pointed out in Reference [21], the typical biological rotary motor is about 8 nm in diameter, 14 nm in length and is capable of producing 20–100 pN·nm of rotary torque. Its operation depends on the availability of ATP in a suitable aqueous environment.

3.2.4
Ion Channels, the Nanotransistors of Biology

The smallest forms of life are bacteria, which are single cells of micrometer size. Cells are enclosed by an impermeable lipid bilayer membrane, the cell wall. This hydrophobic layer is akin to a soap bubble. Lipid cell walls are ubiquitous in all forms of life. Communication from the cell to the extracellular environment is accomplished in part by ion channels, which allow specific ion species to enter or leave the cell.

Two specific types of transmembrane protein ion channels are the Ca^{++} gated potassium channel, and the voltage-gated potassium ion channel, which is essential to the generation of nerve impulses. The dimensions of these transmembrane proteins are on the same order as their close relatives, the rotary engines, which were characterized as 8 nm in diameter and 14 nm in length.

These ion channel structures are "highly conserved", meaning that the essential units which appeared about 1 billion years ago in single cells, have been elaborated upon, but not essentially changed, in the many different cellular applications that have since evolved.

Ca^{++} gated potassium channel
A model for the Ca^{++} gated potassium channel is shown in Figure 3.5, after Schumacher and Adelman [22] based on [23] and [24]. Here the shaded horizontal slabs represent are the upper (outside) and lower (inside) cell walls. The envisioned ion channel, as closed, is shown on the left (a). The structure, which extends across the cell membrane, has an upper "selectivity filter" (which passes only potassium ions), a central cavity, and a Ca^{++}-controlled gate (lower) which opens the channel (right view, b). In this condition the channel will pass potassium ions, but not other ions. The nanometer scale dimensions of the ancient molecular structure that is modeled in Figure 3.5 are similar to those of the rotary motors shown above.

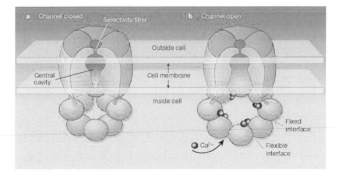

Figure 3.5 Model for Ca^{++}-gated K channel, after [22]

Voltage-gated potassium channel

The long-standing interest in the voltage-gated K$^+$ channel, stems in part from the discovery of Hodgkin and Huxley [25], that nerve impulses are generated by the flux of ions across the lipid membrane of a neuron. These ion channels are responsible for bringing a nerve impulse to an end so a neuron can prepare to fire again. This unit is key to the operation of the central nervous system. It is clear that the voltage difference across the cell membrane serves to open and shut the channel to the flow of potassium ions.

Two competing contemporary models for this channel are depicted in Figure 3.6 [26,27]. In the currently accepted conventional model (upper) the open pore configuration (left) is closed by the lifting of positively cylindrical (helical) units upward by the electric potential difference. In the competing model (lower), closing of the pore to potassium flow results from rotation of "paddle" structures in response to the potential difference.

In both cases the dimensions are nanometer scale and flows of electrical currents are controlled by potential differences. These are the functions of a transistor in elec-

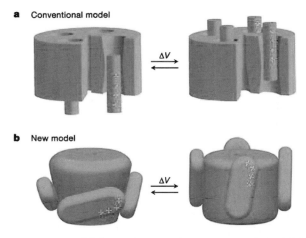

Figure 3.6 Models for a voltage-gated potassium ion channel. [26]

tronics. These ancient biological devices have dimension typically 10 nm, and are thus nanoscale natural versions of transistors.

3.3
How Small can you Make it?

The fundamental limits to the sizes of machines and devices, presented by the size of atoms and as represented by the molecular machines of biology, are clear. The ability within present technology to actually make machines and devices is very limited, so that practical limits on machines and devices are many orders of magnitude larger in size. (There are some notable exceptions: really atomic sized assemblies have been made using the scanning tunneling microscope.) The conventional machine tools that make small mechanical parts scarcely work below millimeter size at present.

Making many identical small molecules, or even many very large molecules is easy for today's chemist and chemical engineer. The challenge of nanotechnology is much harder, to engineer (design and make to order) a complex structure out of molecular sized components.

3.3.1
What are the Methods for Making Small Objects?

The microelectronics process is able to make complex planar structures containing millions of working components in a square centimeter. This process is adaptable to making essentially planar mechanical machines with components on a micrometer scale.

An example of a turbine wheel is shown in Figure 3.7 [28]. This radial inflow turbine is made from silicon, using deep reactive ion etching. It is projected that a hydrocarbon burning gas turbine-electric generator under $1\,cm^3$ in volume could be

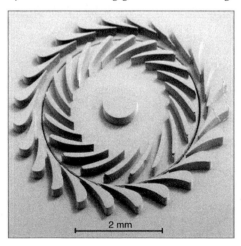

2 mm

Figure 3.7 Turbine wheel produced on silicon wafer with deep reactive ion etching. [28]

built elaborating on this approach, generating up to 50 watts, at an energy density 10 – 30 times that of the most advanced battery materials [28].

3.3.2
How Can you See What you Want to Make?

The motors of biology and the transistors of microelectronics are invisible. Using the tip of a Scanning Tunneling Microscope (STM) as an assembly tool, allows the product to be imaged at the same time. In principle this is a way of making some kinds of nanostructures.

Figure 3.8, after [29] shows steps in the assembly of a ring of iron atoms on a 111 atomically clean surface of copper. In this case the iron atoms are not chemically bonded to the copper, but remain in fixed positions because the equilibrium thermal energy is extremely low in the ultra-high-vacuum cryogenic environment. The Fe atoms actually rest in fixed positions between nearest-neighbor Cu atoms on the 111 Cu surface. The tip of the STM is used to nudge the atoms gently from one of these depressions to a neighboring one, thus assembling the ring. The tip is not used to carry an Fe atom: if it were, the property of the tip for also providing an image would be seriously disturbed.

The interaction of the tip to move the Fe is likely to involve the divergent electric field from the tip, which can induce an electric dipole moment,

$$\mathbf{p} = \alpha \mathbf{E}, \tag{3.7}$$

in the electron cloud of the atom. There is then a dipole interaction energy,

$$U = -\mathbf{p}.\mathbf{E}. \tag{3.8}$$

Because the E field is stronger close to the tip, the dipole is pulled closer to the tip:

$$F = -\mathrm{d}U/\mathrm{d}z = p_z \mathrm{d}E_z/\mathrm{d}z. \tag{3.9}$$

The strength of the force depends on the electric field approximately $E = V/d$, where V is the tip bias voltage and d is its spacing above the surface. Both V and d can be adjusted through the controls of the STM. With the upward force suitably adjusted, it is found that the Fe atom will follow a horizontal displacement of the tip, thus moving the atom. Other forces may contribute here, possibly related to the tunneling current.

This situation therefore does allow a variable force of attraction between the tip and an atom. However, if the atom actually jumps onto the surface of the tip, then further control of that atom is lost.

This situation is a prototype for the imagined molecular assembler tip. It should be realized that STM tips in practice have radii very much larger than an atomic radius. So there is no way that a carried atom could be placed into a recessed position, because access would be blocked by the large radius of the tip. (Tips of smaller radius are found to vibrate excessively.)

Incidentally, the circular ripples, peaking at the center of the circle of Fe atoms are evidence for the wave nature of the electrons in the (111) surface of the Cu sam-

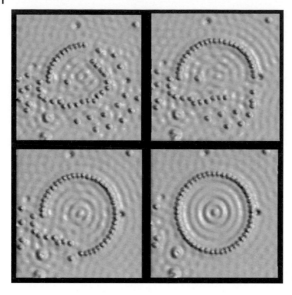

Figure 3.8 Assembling a ring of 48 Fe atoms on a (111) Cu surface with an STM [29]. The diameter of the ring is 14.3 nm

ple. These standing waves result from reflection of the electron waves from the barrier represented by the row of Fe atoms. This situation resembles the reflection of water waves from a solid surface such as a pier or the edge of a canal. The researchers [29] found that the spacing of the crests of the waves was in agreement with the nanophysics of electron waves which will be taken up in Chapter 4. From the nanophysics point of view these ripples represent electrons, acting independently, caught in a two-dimensional circular potential well.

A second example of assembly of a molecular scale object using an STM tip is shown in Figure 3.8 ([30] after Hopkinson, Lutz and Eigler). Here is a grouping of 8 cesium and 8 iodine atoms which have formed a molecule on the (111) surface of copper. It appears that the strong ionic bond of Cs and I has dominated the arrangement, illustrating that an STM tip may not dictate the individual locations of atoms in a structure that is being assembled.

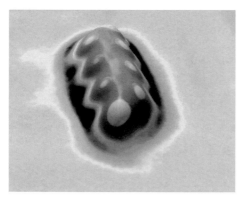

Figure 3.9 Cesium and iodine on Cu 111 [30] This pattern represents a molecule on the copper surface which contains eight cesium and eight iodine atoms. This image illustrates that the assembled atoms may choose their own structure, beyond control of the tip, in this case driven by the strong ionic bond

In this situation one can speculate that the strongest interactions are between Cs and I ions, leading to eight molecules. These molecules are polar and probably arrange themselves in a structure dominated by the dipole-dipole forces between the CsI molecules.

In this case it not clear how strongly the molecules are attached to the copper surface.

3.3.3
How Can you Connect it to the Outside World?

A complex machine to be useful may need multiple connections to its outer environment. An example is the array of wires attached to the edges of a computer chip in a computer. If the premise is that the engineered nanomachine is to perform a function useful to the human-scale world, a large number of connections scaling up from the nanoscale to the centimeter scale may be needed. At the same time, the wiring must be such that unwanted signals or noise from the outer world do not propagate back into the nanoscale device to destroy it.

3.3.4
If you Can't See it or Connect to it, Can you Make it Self-assemble and Work on its Own?

The genius of biology is that complex structures assemble and operate autonomously, or nearly so. Self-assembly of a *complex* nano-structure is completely beyond present engineering approaches, but the example of DNA-directed assembly in biology is understood as an example of what is possible.

3.3.5
Approaches to Assembly of Small Three-dimensional Objects

All of the small objects mentioned so far have been essentially planar. Computer chips have several layers of wiring built up sequentially, but they are essentially planar. The same is true of the MEMS (microelectromechanical) devices based on the semiconductor technology. There are many uses for such devices, of course, but the impact of a fully three-dimensional approach would be very positive.

This situation has been lucidly stated by G. M. Whitesides [31] "The fabrication of microstructures is one of the most pervasive of modern technologies. Almost all microfabrication is now based on photolithography and its dependent technologies, and the dominance of this family of technologies is genuinely remarkable. Photolithography is intrinsically planar, although it can, with difficulty, be induced to produce certain types of nonplanar structures. The development of flexible, economical methods that would have the power of lithography, but would build 3D microstructures, would open the door to a host of applications in microfluidic systems, MEMS, optical devices, and structural systems...."

Variable thickness electroplating

A pioneering approach to making small 3D structures is summarized by Angus and Landau [31]. The approach [32] involves electroplating, with variable thickness images enabled by exposing a photo-sensitized gel through a gray-scale mask, which serves to cross-link the gelatin in proportion to the exposure [31]. The resistance to ionic transport through the gelatin increases with the cross-linking. "Therefore, on electroplating through the gelatin, the gray scale of the original optical mask is translated into thickness variations on the final surface; that is, darker areas on the optical mask lead to thicker electrodeposits. The method provides a convenient additive method for generating 3D surface relief".

Lithography onto curved surfaces

A recent three-dimensional method is summarized in Figure 3.10 [33].

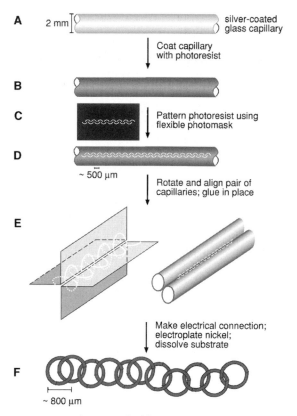

Figure 3.10 Scheme [33] for fabricating a chain using a flexible photomask and electrochemical welding. (A) Metallized glass capillaries coated with photoresist, by pulling them slowly from bulk solution. (B) Capillaries hard-baked at 105 °C for 3 min. Exposure (8s) of coated capillary to UV light through a flexible mask (design shown in (C)) wrapped around its surface. (D) Under optical microscope align two patterned capillaries to be in close proximity with their patterns matched to form a chain. (E) (Links correspond to openings in the photoresist. Dotted lines represent links on the undersides of the capillaries that are not visible from the top). Electroplating nickel for 30 min. at density 20 mA/cm^2 welded together the ends of the chain links, in those areas defined by the photoresist. Finally, release the nickel chain by dissolving photoresist in acetone, dissolving silver metallization in aqueous ferricyanide bath, and dissolving the titanium and glass in concentrated HF [33]

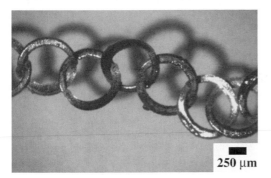

Figure 3.11 Optical micrograph of a free-jointed nickel chain formed by the process shown in Figure 3.10 [33]. The final thickness of the nickel was about 50 μm

This method depends upon a flexible mask, that can be curved around the glass capillary tubes. The size of objects that can be made by this method is limited in part by the relative alignment of the two cylinders that is needed. Generalizations are discussed in [33].

The authors suggest that applications of such fully three-dimensional structures could include ultralight structures for micro air and space vehicles, components for microelectromechanical systems, 3D metallic membranes and electrodes, and, at smaller dimensions, dielectric structures for photonic band gap materials.

Optical tweezers
"Optical tweezers" is the name associated with the use of a focused laser beam to trap and pull a dielectric particle. The effect is somewhat similar to that described in equations (3.7–3.9), but the inhomogeneous electric field is at optical frequency and is most intense at the focus ("waist") of the laser beam. The optical tweezer is typically applied to dielectric spheres, such as polystyrene, of about 1 μm size, which are chemically attached to some biological system of interest. The laser power can be adjusted so that the force on the sphere is appreciable, while the force on single molecules, which are smaller, is negligible. The dielectric sphere will be stably positioned at the center of the trap (waist of the focused laser beam) and as it is displaced from the center, a restoring force proportional to the displacement will be exerted by the trap. An application of this biological technique is indicated in Figure 3.12 [34]. In the left panel (A) the bead, attached to the free DNA strand (template) by the RNA polymerase engine in its copying function, is shown at succeeding times t_1, t_1, t_3. In panel (B) the increasing length of the RNA copy, calibrated in nucleotides (nt), is plotted vs. time [34].

The enzyme RNA polymerase (RNAP) transcribes a DNA template into messenger RNA. In doing so it moves like an engine along the DNA template. (In the experiment shown, the RNAP engine is fixed to the glass slide, and the DNA is pulled to the left through it, in the process of making the copy (RNA, shown as coiling upward in (A) of Figure 3.12). In the experiment the trap position was fixed, and

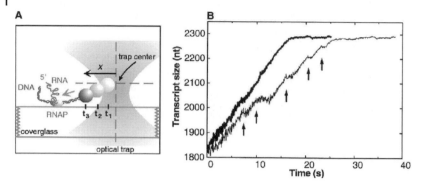

Figure 3.12 Use of optical tweezers to observe single RNAP molecule pulling DNA [34]

the motion of the bead was monitored by an optical interferometer. As the motion proceeded, the restoring force of the trap increased, and at positions marked by arrows in Figure 3.12 (B) the motor stalled, allowing a measurement of the maximum force available. The forces are in the range 21 – 27 pN. After stalling, the trap was repositioned and transcription continued. The upper trace in (B) was constructed by removing the stall phenomena artificially, to determine (from this graph) a replication rate of about 26 nucleotides per second.

Arrays of optical traps

A sophisticated scheme of interfering focused laser light beams has allowed creation and manipulation of 3D optically trapped structures [35]. Figure 3.13 schematically shows the nature of this optical trap and some of the configurations of dielectric spheres that have been stabilized. In Fig 3.13 (A) a detail schematic of the waist of the four-fold beam, which is shown stabilizing an array of eight dielectric spheres. (B) shows six other arrangements of identical dielectric spheres which were stably trapped in this device [35].

These traps are generally useful for micrometer sized particles, as far as individual positioning is concerned. Many particles can be loaded into such traps, encouraging them to condense in a self-organizing fashion. Arrays of more than four traps seem possible, based upon splitting a coherent light beam up into parallel beams.

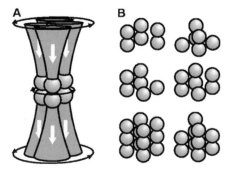

Figure 3:13: Four-fold rotating optical trap stabilizes 3D arrays. [35]

References

[1] M. Planck, Annalen der Physik **4**, 553 (1901).

[2] A. Einstein, Annalen der Physik **17**, 132 (1905).

[3] R. A. Millikan, Physical Review **32**, 349 (1911).

[4] L. Pauling and E. B. Wilson, Jr., *Introduction to Quantum Mechanics with applications to Chemistry*, (Dover, Mineola, N.Y., 1985).

[5] F. L. Pilar, *Elementary Quantum Chemistry*, (Dover, Mineola, N.Y., 2001).

[6] Reprinted with permission from L. Mahadevan and P. Matsudaira, Science **288**, 95 (2000). Copyright 2000 AAAS.

[7] H. Stebbings and J. S. Hyams, *Cell Motility*, (Longman, London, 1979).

[8] Reprinted with permission from R. D. Vale and R. A. Milligan, Science **288**, 88 (2000). Copyright 2000 AAAS.

[9] V. C. Abraham, V. Krishnamurthi, D. L. Taylor, and F. Lanni, Biophys. J. **77**, 1721 (1999).

[10] M. Dogterom and B. Yurke, Science **278**, 856 (1997).

[11] J. E. Molloy, J. E. Burns, J. Kendrick-Jones, R. T. Treager, and D. C. White, Nature **378**, 209 (1995).

[12] S. M. Block, Cell **93**, 5 (1998).

[13] W. B. Amos, Nature **229**, 127 (1971).

[14] R. Niklas, Annu. Rev. Biophys. Biophys. Chem. **17**, 431 (1988).

[15] D. J. DeRosier, Cell **93**, 17 (1998).

[16] R. M. Berry and J. P. Armitage, Adv. Microb. Physiol. **41**, 291 (1999).

[17] J. Tilney and S. Inoue, J. Cell. Biol. **93**, 820 (1982).

[18] A. Yildiz, J. N. Forkey, S. A. McKinney, T. Ha, Y. E. Goldman, and P. R. Selvin, Science **300**, 2061 (2003).

[19] J.E. Molloy and C. Veigel, Science **300**, 2045 (2003).

[20] Reprinted with permission from Y. Sambongi, U. Iko, M. Tanabe, H. Omote, A. Iwamoto-Kihara, I. Ueda, T. Yanagida, Y. Wada, and M. Futai, Science **286**, 1722 (1999). Copyright 1999 AAAS.

[21] Reprinted with permission from R. K. Soong, G. D. Bachand, H. P. Neves, A. G. Olkhovets, H. G. Craighead, and C. D. Montemagno, Science **290**, 1555 (2000). Copyright 2000 AAAS.

[22] Reprinted with permission from Nature: M. Schumacher and J. P.Adelman, Nature **417**, 501 (2002). Copyright 2002, Macmillan Publishers Ltd.

[23] Y. Jiang, A. Lee, J. Chen, M. Cadene, B. T. Chait, and R. MacKinnon, Nature **417**, 515 (2002).

[24] Y. Jiang, A. Lee, J. Chen, M. Cadene, B. T. Chait, and R. MacKinnon, Nature **417**, 523 (2002).

[25] A. L. Hodgkin and A. F. Huxley, J. Physiol. (London) **117**, 500 (1952).

[26] Reprinted with permission from Nature: Y. Jiang, A. Lee, J. Chen, V. Ruta, M. Cadene, B. T. Chait, and R. MacKinnon, Nature **423**, 33 (2003). Copyright 2003, Macmillan Publishers Ltd.

[27] Y. Jiang, V. Ruta, J. Chen, A. Lee, and R. MacKinnon, Nature **423**, 42 (2003).

[28] Courtesy Martin A. Schmidt, Microsystems Technology Laboratories at Massachusetts Institute of Technology.

[29] Courtesy IBM Research, Almaden Research Center. Unauthorized use not permitted.

[30] Courtesy IBM Research, Almaden Research Center. Unauthorized use not permitted.

[31] J. C. Angus, U. Landau, and G. M. Whitesides, Science **281**, 1143 (1998).

[32] J. C. Angus, U. Landau, S. H. Liao, and M. C. Yang, J. Electrochem. Soc. **133**, 1152 (1986).

[33] Reprinted with permission from R. J. Jackman, S. T. Brittain, A. Adams, M. G. Prentiss and G. M. Whitesides, Science **280**, 2089–2091 (1998). Copyright 1998 AAAS.

[34] Reprinted with permission from M. D. Wang *et al.*, Science **282**, 902–907 (1998). Copyright 1998 AAAS.

[35] Reprinted with permission from M. P. MacDonald, L. Paterson, K. Volke-Sepulveda, J. Arit, W. Sibbett, and K. Dholakia, Science **296**, 1101–1103 (2002). Copyright 2002 AAAS.

4
Quantum Nature of the Nanoworld

The particles of matter (electrons, protons, neutrons) provide limits to the smallness of anything composed of chemical matter, which is itself composed of atoms. The rules that these particles obey are different from the rules of macroscopic matter. An understanding of the rules is useful to understand the structure of atoms and chemical matter. An understanding of these rules, and of the relation between the wave and particle natures of light, is also key to understanding deviations from behavior of devices and machines that will appear as their dimensions are reduced toward the atomic size in any process of scaling.

One of the areas in which anticipated failures of classical scaling are of intense interest is in microelectronic devices, where it is obvious that Moore's Law (Figure 1.2), of increasing silicon chip performance, will eventually be modified. For one thing, the new rules allow a particle to penetrate, or tunnel through, a barrier. Such an effect is becoming more likely in the gate oxide of field effect transistors, as scaling efforts reduce its thickness.

The other aspect of the new nanophysical rules is that they permit new device concepts, such as the Esaki tunnel diode, mentioned in Chapter 1.

The new rules say that all matter is made of particles, but that the particles have an associated wave property. That wave property was noted in connection with Figure 3.8, which exhibited electron ripples.

One of the first quantum rules discovered, which had a large impact on understanding of atoms and their absorption and emission of light, was the quantization of angular momentum and energy levels of an electron in a hydrogen atom. This was first discovered in 1913 by Nils Bohr [1], whose work also stimulated further progress until by 1926 or so a more complete understanding of the wave properties of matter was achieved by Schrodinger [2].

Nanophysics and Nanotechnology: An Introduction to Modern Concepts in Nanoscience. Edward L. Wolf
Copyright © 2004 WILEY-VCH Verlag GmbH & Co. KGaA, Weinheim
ISBN: 3-527-40407-4

4.1
Bohr's Model of the Nuclear Atom

The structure of the atom is completely nanophysical, requiring quantum mechanics for its description. Bohr's semi-classical model of the atom was a giant step toward this understanding, and still provides much useful information. By introducing, in 1913, a completely arbitrary "quantum number", Bohr [1] was able to break the long-standing failure to understand how an atom could have sharply defined energy levels. These levels were suggested by the optical spectra, which were composed of sharp lines.

Bohr's model is planetary in nature, with the electron circling the nucleus. The model was based on information obtained earlier: that the nucleus of the atom was a tiny object, much smaller in size than the atom itself, containing positive charge Ze, with Z the atomic number, and e the electron charge, 1.6×10^{-19} C. The nucleus is much more massive than the electron, so that its motion will be neglected.

Bohr's model describes a single electron orbiting a massive nucleus of charge $+Ze$. The attractive Coulomb force $F = kZe^2/r^2$, where $k = (4\pi\varepsilon_0)^{-1} = 9 \times 10^9 \, \text{Nm}^2/\text{C}^2$, balances $m_e v^2/r$, which is the mass of the electron, $m_e = 9.1 \times 10^{-31}$ kg, times the required acceleration to the center, v^2/r. The total energy of the motion, $E = mv^2/2 - kZe^2/r$, adds up to $-kZe^2/2r$. This is true because the kinetic energy is always -0.5 times the (negative) potential energy in a circular orbit, as can be deduced from the mentioned force balance.

There is thus a crucial relation between the total energy of the electron in the orbit, E, and the radius of the orbit, r.

$$E = -kZe^2/2r. \tag{4.1}$$

This classical relation predicts collapse (of atoms, of all matter): for small r the energy is increasingly favorable (negative). So the classical electron will spiral in toward $r = 0$, giving off energy in the form of electromagnetic radiation. All chemical matter is unstable to collapse in this firm prediction of classical physics.

4.1.1
Quantization of Angular Momentum

Bohr, recognizing that such a collapse does not occur, was emboldened to impose an arbitrary quantum condition to stabilize his model of the atom. Bohr's postulate of 1913 was of the quantization of the angular momentum L of the electron of mass m circling the nucleus, in an orbit of radius r and speed v:

$$L = mvr = n\hbar = nh/2\pi. \tag{4.2}$$

Here n is the arbitrary integer quantum number $n = 1,2\ldots$. Note that the units of Planck's constant, $J \cdot s$, are also the units of angular momentum. This additional constraint leads easily to the basic and confirmed properties of the "Bohr orbits" of electrons in hydrogen and similar one-electron atoms:

$$E_n = -kZe^2/2r_n, \quad r_n = n^2 a_o/Z, \quad \text{where } a_o = \hbar^2/mke^2 = 0.053 \text{ nm.} \tag{4.3}$$

Here, and elsewhere, k is used as a shorthand symbol for the Coulomb constant $k = (4\pi\varepsilon_o)^{-1}$.

The energy of the electron in the nth orbit can thus be given as $E_n = -E_o Z^2/n^2$, $n = 1,2,\ldots$, where

$$E_o = mk^2e^4/2\hbar^2 = 13.6 \text{ eV.} \tag{4.3a}$$

All of the spectroscopic observations of anomalous discrete light emissions and light absorptions of the one-electron atom were nicely predicted by the simple quantum condition

$$h\nu = hc/\lambda = E_o(1/n_1^2 - 1/n_2^2). \tag{4.4}$$

The energy of the light is exactly the difference of the energy of two electron states, n_1, n_2 in the atom. This was a breakthrough in the understanding of atoms, and stimulated work toward a more complete theory of nanophysics which was provided by Schrodinger in 1926 [2].

The Bohr model, which does not incorporate the basic wavelike nature of microscopic matter, fails to precisely predict some aspects of the motion and location of electrons. (It is found that the idea of an electron orbit, in the planetary sense, is wrong, in nanophysics.)

In spite of this, the electron energies $E_n = -E_o Z^2/n^2$, spectral line wavelengths, and the characteristic size of the electron motion, $a_o = \hbar^2/mke^2 = 0.053$ nm, are all exactly preserved in the fully correct treatment based on nanophysics, to be described below.

4.1.2
Extensions of Bohr's Model

The Bohr model also remains useful in predicting the properties of "hydrogenic" electrons bound to donor impurity ions in semiconductors. This analysis explains how the carrier concentrations and electrical conductivities of industrial semiconductors are related to the intentionally introduced donor and acceptor impurity concentrations, N_D and N_A, respectively. The additional ideas needed are of the relative dielectric constant of the semiconductor and the "effective mass" that an electron exhibits as it moves in a semiconductor.

The Bohr model is also useful in analyzing the optical spectra of semiconductors exposed to radiation, having energy $E = hc/\lambda > E_g$ which produces electron–hole pairs. (E_g is the symbol for the "energy gap" of a semiconductor, which is typically about 1 eV.) It is found that such electrons and holes, attracted by the Coulomb force, momentarily orbit around each other, described by the mathematics of the Bohr model, and emit photons whose energies are predicted by the relevant Bohr model. Such states, called "excitons", are well documented in experiments measuring the spectra of fluorescent light from optically irradiated semiconductors.

A relevant topic in nanophysics is the alteration, from the exciton spectrum, of the fluorescent light emitted by a semiconductor particle as its size, L, is reduced. It is found that the correct light emission wavelengths for small sample sizes L, are obtained from the energies of electrons and holes contained in three-dimensional potentials, using the Schrodinger equation. This understanding is the basis for the behavior of "quantum dots", marketed as fluorescent markers in biological experiments, as will be described below.

4.2
Particle-wave Nature of Light and Matter, DeBroglie Formulas $\lambda = h/p$, $E = h\nu$

One of the most direct indications of the wave nature of light is the sinusoidal interference pattern of coherent light falling on a screen behind two linear slits of small spacing, d. The rule for appearance of maxima at angular position θ in the interference pattern,

$$n\lambda = d\sin\theta, \tag{4.5}$$

is that the difference in the path length of the light from the two slits shall be an integral number n of light wavelengths, $n\lambda$. Dark regions in the interference pattern occur at locations where the light waves from the two slits arrive 180 degrees out of phase, so that they exactly cancel.

The first prediction of a wave nature of matter was given by Louis DeBroglie [3]. This young physics student postulated that since light, historically considered to be wavelike, was established to have a particle nature, it might be that matter, considered to be made of particles, might have a wave nature. The appropriate wavelength for matter, DeBroglie suggested, is

$$\lambda = h/p, \tag{4.6}$$

where h is Planck's constant, and $p = mv$ is the momentum. For light $p = E/c$, so the relation $\lambda = h/p$ can be read as $\lambda = hc/E = c/\nu$. Filling out his vision of the symmetry between light and matter, DeBroglie also said that the frequency ν associated with matter is given by the same relation,

$$E = h\nu, \tag{4.7}$$

as had been established for light by Planck.

This postulated wave property of matter was confirmed by observation of electron diffraction by Davisson and Germer [4]. The details of the observed diffraction patterns could be fitted if the wavelength of the electrons was exactly given by h/p. For a classical particle, $\lambda = h/(2mE)^{1/2}$, since $p^2 = 2mE$. So there was no doubt that a wave nature for matter particles is correct, as suggested by DeBroglie [3]. The question then became one of finding an equation to determine the wave properties in a given situation.

4.3
Wavefunction Ψ for Electron, Probability Density Ψ*Ψ, Traveling and Standing Waves

The behavior of atomic scale particles is guided by a wavefunction, $\Psi(r,t)$, which is usually a complex number. The probability of finding the particle is given by the square of the absolute value

$$\Psi^*(r,t)\Psi(r,t) = P(r,t) \tag{4.8}$$

of the wavefunction. This quantity P is a probability density, so that the chance of finding the particle in a particular small region $dxdydz$ is $Pdxdydz$.

Complex Numbers x+iy. The complex number is a notation for a point in the xy plane, where the symbol "i" acts like a unit vector in the y direction, formally obtained by rotating a unit vector along the x-axis in the ccw direction by $\pi/2$ radians. Thus $i^2 = -1$. The *complex conjugate* of the complex number, $x + iy$, is obtained by changing the sign of y, and is thus $x-iy$. The absolute value of the complex number is the distance r from the origin to the point x,y, namely

$$r = (x^2 + y^2)^{1/2} = [(x + iy)(x-iy)]^{1/2}. \tag{4.9}$$

A convenient representation of a complex number is

$$r\exp(i\theta) = r(\cos\theta + i\sin\theta), \text{ where } \theta = \tan^{-1}(y/x). \tag{4.10}$$

The wavefunction should be chosen so that P is normalized. That is,

$$\iiint P(x,y,z)dxdydz = 1 \tag{4.11}$$

if the integral covers the whole region where the electron or other particle may possibly exist.

There are other properties that a suitable wavefunction must have, as we will later discuss. Similarly, a many-particle $\Psi(r_1,r_2,\ldots,r_n,t)$ and probability $P(r_1,r_2,\ldots,r_n,t)$ can be defined.

The wavefunction for a beam of particles of identical energy $p^2/2m$ in one dimension is a traveling wave

$$\Psi(x,t) = L^{-1/2}\exp(ikx-i\omega t) = L^{-1/2}[\cos(kx-\omega t) + i\sin(kx-\omega t)], \tag{4.12}$$

where $k = 2\pi/\lambda$ and $\omega = 2\pi\nu$. According to the DeBroglie relation (4.6)

$$p = h/\lambda = \hbar k, \tag{4.13}$$

where $\hbar = h/2\pi$. Similarly, from (4.7),

$$\omega = 2\pi\nu = E/\hbar. \tag{4.14}$$

The normalization gives one particle in each length L, along an infinite x-axis. A point of fixed phase (such as a peak in the real part of the wave) moves as $x = (\omega/k)t$, so (ω/k) is called the phase velocity

$$v_{ph} = (\omega/k) \tag{4.15}$$

of the wave. Note that

$$\Psi(x,t) = L^{-1/2}e^{(ikx - i\omega t)} = L^{-1/2}\exp(ikx - i\omega t) \tag{4.16}$$

has a constant absolute value at any x, describing a particle equally likely to be at any position on the infinite x-axis. There is no localization in this wavefunction since the momentum is perfectly described, implying $\Delta x = \infty$.

4.4
Maxwell's Equations; *E* and *B* as Wavefunctions for Photons, Optical Fiber Modes

The laws of electricity and magnetism give the values of electric field E and magnetic field B, as functions of position, in various circumstances. In classical electricity and magnetism it is known that the energy density in the electromagnetic field is

$$(\varepsilon_o E^2 + \mu_o B^2)/2. \tag{4.17}$$

Since the electromagnetic energy resides in particles called photons, this classical energy density can also be thought of as a probability function for finding photons. The analogy is most direct if the E and B fields represent traveling waves, and thus transport energy.

The optical fiber presents a situation for the electric field E in which the analogy with the wavefunction of quantum mechanics is clear. An optical fiber consists of a cylindrical quartz glass core with index of refraction n_1 contained within concentric silica glass cladding of somewhat smaller index n_2. Typical radii for a single mode fiber are $5 - 10\,\mu m$ for the inner core and $120\,\mu m$ for the cladding. For light traveling along the core, parallel, or nearly parallel, to its axis, total internal reflection occurs, confining the light to the core. The critical angle for total internal reflection is given by

$$\theta_r = \sin^{-1}(n_2/n_1). \tag{4.18}$$

Note that this angle is measured from normal incidence. The light proceeds down the core of the optical fiber as a transverse electric wave. The radial distribution of transverse electric field E (perpendicular to the axis) in the symmetric mode (the only mode possible in the smallest, single mode, fibers) is a smooth function peaked at the axis, $r = 0$, and falling in a bell shaped function with increasing radius.

The electric field only weakly extends into the cladding region, where it decays exponentially with increasing radius. This is a behavior qualitatively similar to that of a wave function in quantum mechanics representing a particle whose energy is less than the potential energy.

4.5
The Heisenberg Uncertainty Principle

The uncertainty principle [5] is a consequence of a wave description of the location of a particle. It states that the position x and the momentum p of a particle can both be simultaneously known only to minimum levels of uncertainty, Δx and Δp, respectively, where

$$\Delta x \, \Delta p \geq \hbar/2. \tag{4.19}$$

The free particle of precise momentum $p = \hbar k$, described by the wavefunction $\Psi(x,t) = L^{-1/2}\exp(ikx-i\omega t)$ (4.16) represents a case of $\Delta x = \infty$ and $\Delta p \cong 0$.

A wavefunction that describes a localized particle can be constructed by adding waves having a distribution of k and ω values. An example of such a linear combiniation, using a trigonometric identity, is

$$\Psi = A[\cos(k_1 x-\omega_1 t) + \cos(k_2 x-\omega_2 t)] = 2A\cos[(1/2)\Delta kx-(1/2)\Delta\omega t]\cos(k_{av}x-\omega_{av}t), \tag{4.20}$$

where $\Delta k = k_2-k_1$, $\Delta\omega = \omega_2-\omega_1$, $k_{av} = (k_2 + k_1)/2$, and $\omega_{av} = (\omega_2 + \omega_1)/2$.

Assuming the two k's and ω's are closely spaced, so that $\Delta k < k_{av}$ and $\Delta\omega < \omega_{av}$, then the $\cos[(1/2)\Delta kx-(1/2)\Delta\omega t]$ factor acts like an *envelope function* of long wavelength and low frequency which modulates the original wave. At fixed t, the spacing Δx between adjacent nodes of $\cos(\Delta kx/2)$ is

$$\Delta x = 2\pi/\Delta k. \tag{4.21}$$

This width can be considered as a length of localization for the particle represented by the superposition of two waves. The equivalent relation $\Delta k\Delta x = 2\pi$ is a form of the uncertainty principle, which we see is simply a wave property. Since $\Delta p = \hbar\Delta k$, from the DeBroglie relation, this gives $\Delta p\Delta x = h$. A tighter localization, achieved by a wavepacket composed of a Gaussian distribution of k values, gives the smaller uncertainty product quoted above, $\Delta x \, \Delta p \geq \hbar/2$.

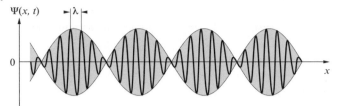

$\Psi(x,\, t)$

0

x

Figure 4.1 Adding waves creates regions of localization that move at the group velocity

Finally, the motion of the envelope function in (4.20) is described by $x = (\Delta\omega/\Delta k)t$. The new velocity is called the group velocity

$$v_\mathrm{g} = (\partial\omega/\partial k). \tag{4.22}$$

To apply this simple analysis to the motion of a localized particle of energy $E = \hbar\omega = mv^2/2m = \hbar^2 k^2/2m$, let us calculate the expected classical velocity. From the foregoing, $\omega = \hbar k^2/2m$, so $\partial\omega/\partial k = \hbar k/m = v$. Thus, the group velocity $v = \partial\omega/\partial k$, and DeBroglie's relations correctly reproduce the classical velocity p/m of a particle.

4.6
Schrodinger Equation, Quantum States and Energies, Barrier Tunneling

The equation to provide the wavefunction $\Psi(x,t)$ describing the location of a particle in a given physical situation was provided by Schrodinger [2]. This equation may seem mysterious, but actually it is not so, at least in hindsight. Any successful wave equation has to reflect the DeBroglie matter-wave relations [3], $E = \hbar\omega$ and $p = \hbar k$ ($\lambda = h/p$). The correct equation must provide a traveling wave solution $\Psi(x,t) = L^{-1/2}\exp(ikx - i\omega t)$ for a free particle, in order to match the electron diffraction observations of Davisson and Germer [4].

Further guidance in finding the correct matter-wave equation is afforded by Maxwell's wave equation for all electromagnetic waves, from elementary physics,

$$\partial^2 \Psi(x,t)/\partial x^2 - \varepsilon_\mathrm{o}\mu_\mathrm{o}\,\partial^2 \Psi(x,t)/\partial t^2 = 0. \tag{4.23}$$

This equation, where, for the moment, $\Psi(x,t)$ represents a component (e.g., E_y) of the **E** or **B** vectors, was obtained by combining the experimentally determined laws of electricity and of magnetism. Applying the second derivatives of Maxwell's wave equation (4.23) to the traveling wavefunction $\Psi(x,t) = \exp(ikx - i\omega t)$, we find

$$\partial^2 \Psi(x,t)/\partial x^2 = -k^2 \Psi(x,t),\ \partial^2 \Psi(x,t)/\partial t^2 = -\omega^2 \Psi(x,t). \tag{4.24}$$

Substitution of the relations (4.24) into the equation (4.23) produces the condition

$$[-k^2 + \omega^2 \varepsilon_\mathrm{o}\mu_\mathrm{o}]\,\Psi(x,t) = 0. \tag{4.25}$$

This equality requires $\omega/k = (\varepsilon_0\mu_0)^{-1/2} = 2.99793 \times 10^8 \text{m/s}$. This speed is the measured speed of light, and makes clear then the origin of all electromagnetic waves as similar to light.

The factor in brackets above (4.25) relates to the speed of the light wave. The question now is how to generate a matter wave equation so that an analogous bracket term will give some condition on matter.

4.6.1
Schrodinger Equations in one Dimension

A good guess for the corresponding bracketed factor in the matter wave equation is a statement of the energy of the particle, $K + U = E$, or, using the DeBroglie relations:

$$[\hbar^2 k^2/2m + U - \hbar\omega] \, \Psi(x,t) = 0. \tag{4.26}$$

Based on this correct statement of conservation of energy, and knowing the solution $\Psi(x,t) = \exp(ikx - i\omega t)$, the equation has to involve $\partial^2 \Psi(x,t)/\partial x^2$, as before. In addition, the first time-derivative $\partial \Psi(x,t)/\partial t = -i\omega \, \Psi(x,t)$ is needed, in order to produce the $\hbar\omega$ in the statement of conservation of energy, which in (4.26) appears in the bracket factor.

Time-dependent Equation
On this basis, the Schrodinger equation in the one-dimensional case, with time-dependent potential $U(x,t)$, is

$$-\hbar^2/2m \, \partial^2 \, \Psi(x,t)/\partial x^2 + U(x,t) \, \Psi(x,t) = i\hbar \, \partial \, \Psi(x,t)/\partial t = \mathcal{H} \Psi. \tag{4.27}$$

The left-hand side of the equation is sometimes written $\mathcal{H}\Psi$, with $\mathcal{H}$ the operator which represents the energy terms on the left side of the equation.

Time-independent Equation
In the frequent event that the potential U is time-independent, a product wavefunction

$$\Psi(x,t) = \psi(x)\varphi(t), \tag{4.28}$$

when substituted into the time-dependent equation above, yields

$$\varphi(t) = \exp(-iEt/\hbar), \tag{4.29}$$

and the time-independent Schrodinger equation,

$$-(\hbar^2/2m)d^2\psi(x)/dx^2 + U\psi(x) = E\psi(x), \tag{4.30}$$

to be solved for $\psi(x)$ and energy E. The solution $\psi(x)$ must satisfy the equation and also boundary conditions, as well as physical requirements.

The physical requirements are that $\psi(x)$ be continuous, and have a continuous derivative except in cases where the U is infinite. A second requirement is that the integral of $\psi^*(x)\psi(x)$ over the whole range of x must be finite, so that a normalization can be found. This may mean, for example, in cases where real exponential solutions $\exp(\kappa x)$, $\exp(-\kappa x)$ satisfy the equation, that the positive exponential solution can be rejected as non-physical.

The momentum of a particle has been associated in our treatment with $-i\hbar\partial/\partial x$, which when operated on $\exp(ikx)$ gives the momentum $\hbar k$ times $\exp(ikx)$. One can see from this that a real, as opposed to a complex, wavefunction, will not represent a particle with real momentum.

4.6.2
The Trapped Particle in one Dimension

The simplest problem is a trapped particle in one dimension. Suppose $U = 0$ for $0 < x < L$, and $U = \infty$ elsewhere, where $\psi(x) = 0$. For $0 < x < L$, the equation becomes

$$d^2\psi(x)/dx^2 + (2mE/\hbar^2)\,\psi(x) = 0. \tag{4.31}$$

This has the same form as the simple harmonic oscillator equation (2.1) discussed earlier, so the solutions (2.2) and (2.3), similarly, can be written as

$$\psi(x) = A\sin kx + B\cos kx, \text{ where} \tag{4.32}$$

$$k = (2mE/\hbar^2)^{1/2} = 2\pi/\lambda. \tag{4.33}$$

The infinite potential walls at $x = 0$ and $x = L$ require $\psi(0) = \psi(L) = 0$, which means that $B = 0$. Again, the boundary condition $\psi(L) = 0 = A\sin kL$ means that

$$kL = n\pi, \quad \text{with } n = 1,2\ldots \tag{4.34}$$

This, in turn, gives

$$E_n = \hbar^2(n\pi/L)^2/2m = n^2h^2/8mL^2, \quad n = 1,2,3,\ldots \tag{4.35}$$

We see that the allowed energies increase as the square of the integer quantum number n, and that the energies increase quadratically as L is decreased.

Finally, the normalization constant A of the wavefunction $\psi_n(x) = A\sin(n\pi x/L)$ must be chosen to make the probability of finding the particle somewhere in the trap to be unity. Thus

$$A^2\int_0^L \sin^2(n\pi x/L)dx = 1 \tag{4.36}$$

implies $A = (2/L)^{1/2}$.

The condition for allowed values of $k = n\pi/L$ is equivalent to

$$L = n\lambda/2, \tag{4.37}$$

the same condition that applies to waves on a violin string.

The exact solution of this simple problem illustrates typical nanophysical behavior in which there are discrete allowed energies and corresponding wavefunctions. The wavefunctions do not precisely locate a particle, they only provide statements on the probability of finding a particle in a given range.

For example, the probability distribution for $n = 3$ is $P_3(x) = (2/L) \sin^2(3\pi x/L)$. This normalized function has three identical peaks on the interval $0, L$, and one can see by inspection that the probability of finding the particle in the range $L/3$ to $2L/3$ is exactly $1/3$. As the quantum number n becomes large, the function $P_n(x)$ approaches the classical probability distribution, $P = 1/L$, as the oscillations of $\sin^2(n\pi x/L)$ become too rapid to observe.

Linear Combinations of Solutions

An additional property of differential equations such as the Schrodinger equation is that linear combinations of solutions are also solutions. So a possible linear combination solution is

$$\Psi_{1+3}(x,t) = A[\sin(\pi x/L) \exp(-iE_1 t/\hbar) + \sin(3\pi x/L) \exp(-iE_3 t/\hbar)]. \tag{4.38}$$

This will lead to a time-dependent probability density.

Expectation Values

The $P(x)$ function can be used to obtain precise expectation values. For example, in the case of the trapped particle in one dimension,

$$<x_n^2> = (2/L) \int_0^L x^2 \sin^2(n\pi x/L) dx = (L^2/3)(1 - \frac{3}{n^2\pi^2}), \quad n = 1, 2... \tag{4.39}$$

This result, exact for any quantum number n, approaches the classical value $(L^2/3)$ as the quantum number n becomes large. This is a general property in quantum mechanics, that the classical result is recovered in the limit of large quantum numbers.

The expectation value of the coordinate, x, between states m, n is

$$<x_{m,n}> = (2/L) \int_0^L x \sin(m\pi x/L) \exp(-iE_m t/\hbar) \sin(n\pi x/L) \exp(iE_n t/\hbar) dx$$

$$= x_{m,n} \exp(-i\omega_{m,n}t), \tag{4.40}$$

where $\omega_{m,n} = (E_m - E_n)/\hbar$.

For a charged particle, $p = e<x_{m,n}>$ represents an oscillating electric dipole moment which can lead to absorption or emission of a photon.

The expectation value of the energy in any particular quantum state is

$$<E> = (2/L) \int_0^L \psi^* \mathcal{H}\, \psi dx. \tag{4.41}$$

Two-particle Wavefunction

The wavefunction for two non-interacting particles simultaneously present in the 1D trap, in quantum states *n,m*, respectively, can be written as

$$\psi_{n,m}(x_1, x_2) = A^2 \sin(n\pi x_1/L) \sin(m\pi x_2/L) = \psi_n(x_1)\, \psi_m(x_2). \tag{4.42}$$

The corresponding probability density is

$$P_{n,m}(x_1, x_2) = \psi^*_{n,m}(x_1, x_2)\, \psi_{n,m}(x_1, x_2). \tag{4.43}$$

4.6.3
Reflection and Tunneling at a Potential Step

A particle moving toward a finite potential step U_o at $x = 0$ illustrates reflection and tunneling effects which are basic features of nanophysics. Suppose $U = 0$ for $x < 0$ and $U = U_o$ for $x \geq 0$.

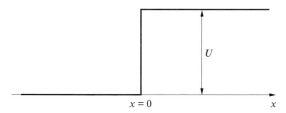

Figure 4.2 Finite potential step at $x = 0$

Case 1: $E > U_o$

In case $E > U_o$, the particle incident from negative x is partially transmitted and partially reflected. For negative x,

$$\psi(x) = A \exp(ikx) + B \exp(-ikx), \text{ where } k = (2mE/\hbar^2)^{1/2}. \tag{4.44}$$

For positive x,

$$\psi(x) = C \exp(ik'x) + D \exp(-ik'x), \text{ where } k' = [(2m(E-U_o)/\hbar^2]^{1/2}. \tag{4.45}$$

Here A, B, C, and D are arbitrary complex numbers. The physical constraints on the allowable solutions are essential to solving this problem.

First, $D = 0$, since no particles are incident from the right.

Second, at $x = 0$ the required continuity of $\psi(x)$ implies $A + B = C$.

Third, at $x = 0$ the derivatives, $d\psi(x)/dx = Aik\exp(ikx) - Bik\exp(-ikx)$ on the left, and $d\psi(x)/dx = Cik'\exp(ik'x)$, on the right, must be equal. Thus

$$A + B = C \quad \text{and} \quad k'C = k(A - B). \tag{4.46}$$

The equations (4.46) are equivalent to

$$B = \frac{k - k'}{k + k'} A \quad \text{and} \quad C = \frac{2k}{k + k'} A. \tag{4.47}$$

The reflection and transmission probabilities, R and T, respectively, for the particle flux are then

$$R = \frac{|B|^2}{|A|^2} = \left(\frac{k - k'}{k + k'}\right)^2 \quad \text{and} \quad T = \frac{k'|C|^2}{k|A|^2} = \frac{4kk'}{(k + k)^2}. \tag{4.48}$$

It is easy to see that equation (4.48) implies

$$R + T = 1. \tag{4.49}$$

These equations are equivalent to the laws for reflection and transmission of light waves at a discontinuity in the refractive index. In the case of particles, it means that a particle is partially reflected even if it has enough energy to propagate past the jump in potential U_o.

Case 2: $E < U_o$

The only change is that now $E - U_o$ is negative, making k' an imaginary number. For this reason k' is now written as $k' = i\kappa$, where

$$\kappa = [2m(U_o - E)/\hbar^2]^{1/2} \tag{4.50}$$

is a real decay constant. Now the solution for positive x becomes

$$\psi(x) = C\exp(-\kappa x) + D\exp(\kappa x), \text{ where } \kappa = [2m(U_o - E)/\hbar^2]^{1/2}. \tag{4.51}$$

In this case, $D = 0$, to prevent the particle from unphysically collecting at large positive x. Equations (4.47) and the first of equations (4.48) remain valid setting $k' = i\kappa$. It is seen that $R = 1$, because the numerator and denominator in the first equation (4.48) are complex conjugates of each other, and thus have the same absolute value.

The solution for positive x is now an exponentially decaying function, and is not automatically zero in the region of negative energy. Find the probability at $x = 0$, $|C|^2$, by setting $k' = i\kappa$ in equation (4.47) and forming $|C|^2 = C^*C$, to be

$$|C|^2 = |A|^2 \frac{4k^2}{(k^2 + \kappa^2)} = |A|^2 \frac{4E}{[E + (U_o - E)]} = |A|^2 \frac{4E}{U_o}, \tag{4.52}$$

where $E = (\hbar^2 k^2/2m) < U_o$. Note that $|C|^2 = 0$ for an infinite potential, as in our treatment of the trapped particle. Also, this expression agrees in the limit $E = U_o$ with equation (4.47).

Thus, the probability of finding the particle in the forbidden region of positive x is

$$P(x>0) = |2A|^2 (E/U_o) \int_0^\infty \exp(-2\kappa x)dx = 2|A|^2 E/(\kappa \ U_o), \tag{4.53}$$

where A is the amplitude of the incoming wave and $E < U_o$.

This is an example of the nanophysical tunneling effect, a particle can be sometimes found where its classical energy is negative, $E < U$.

The electron waves on the left of the barrier, $x < 0$, from (4.44) with $|A| = |B|$, add up to

$$\psi(x) = 2|A| \cos(kx + \delta), \tag{4.54}$$

which is a standing wave. Here the phase shift δ of the standing wave brings equation (4.54), when evaluated at $x = 0$, into agreement with equation (4.52). This standing wave will give "ripples" whenever an electron wave is reflected from a barrier, and the wavelength will be $\lambda = 2\pi/k$, where $k = (2mE/\hbar^2)^{1/2}$, given in (4.44). Actually, the observable quantity is

$$P(x) = \psi^*(x) \ \psi(x) = 4|A|^2 \cos^2(kx + \delta), \tag{4.55}$$

so the observed wavelength will be halved. Such ripples are evident in Figure 3.8.

4.6.4
Penetration of a Barrier

If the problem is turned into a tunneling barrier penetration problem by limiting the region of potential U_o to a width t: $0 < x < t$, the essential dependence of the tunneling transmission probability $|T|^2$ is proportional to $\exp(-2\kappa t)$. In this case there will be a traveling wave, say of amplitude F, for large positive x: $\psi(x) = F \exp(ikx)$, and the relevant transmission probability is defined as the ratio

$$|T|^2 = |F|^2/|A|^2, \tag{4.56}$$

where A is the amplitude of the wave coming from negative x values. To solve this problem the wavefunctions in the three separate regions: $x < 0$, $0 < x < t$, and $x > t$ are written and adjusted to match in value and in slope at $x = 0$ and at $x = t$. The result in case $\kappa t \gg 1$ is

$$|T|^2 = \frac{16k^2\kappa^2\exp(-2\kappa t)}{(k^2+\kappa^2)^2}, \tag{4.57}$$

where the symbols have the same meanings as above, and the exponential function provides the dominant dependence.

4.6.5
Trapped Particles in Two and Three Dimensions: Quantum Dot

The generalization of the Schrodinger equation to three dimensions is

$$-(\hbar^2/2m)\nabla^2\psi(\mathbf{r}) + U(\mathbf{r})\ \psi(\mathbf{r}) = E\psi(\mathbf{r}), \tag{4.58}$$

where $\nabla^2 = \partial^2/\partial x^2 + \partial^2/\partial y^2 + \partial^2/\partial z^2$ and $\mathbf{r}$ is a vector with components x,y,z.

It is not difficult to see that the solutions for a particle in a three-dimensional infinite trap of volume L^3 with impenetrable walls are given as

$$\psi_n(x,y,z) = (2/L)^{3/2}\sin(n_x\pi x/L)\ \sin(n_y\pi y/L)\ \sin(n_z\pi z/L), \quad \text{where } n_x = 1,2\ldots, \text{ etc.,} \tag{4.59}$$

and

$$E_n = [h^2/8mL^2](n_x^2 + n_y^2 + n_z^2). \tag{4.60}$$

These simple results can be easily adapted to two-dimensional boxes and also to boxes of unequal dimensions $L_x\ L_y\ L_z$.

Electrons Trapped in a Two-dimensional Box

A scanning tunneling microscope image, shown in Figure 4.3, reveals some aspects of the trapping of electrons in a 2D rectangular potential well. The well in this case is generated by the rectangular array of iron atoms (silver-colored dots in this image), which reflect electrons, much as in equation (4.55). The electrons are free electrons on the surface of (111) oriented single crystal copper. These electrons, by a quirk of the solid state physics of (111) copper, are essentially confined to the sur-

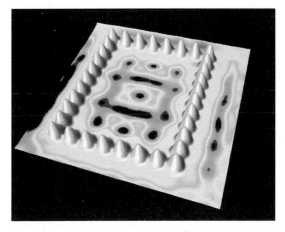

Figure 4.3 Electrons trapped in a small two-dimensional box on the (111) surface of copper. A rectangular array of iron atoms serves as a barrier, reflecting electron waves. [6]

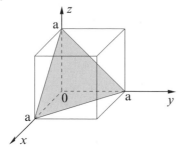

Figure 4.4 Geometry of a (111) plane, shown shaded. Copper is a face-centered cubic crystal, but only if the surface is cut to consist of the indicated (111) plane will the 2D electron effects be present

face, and are further confined by the iron atoms to stay inside (or outside) the box. The traces of "ripples" which are seen here (also seen more clearly in Figure 3.8 in a circular geometry) are visible because the STM measurement, on a specifically 2D electron system, is sensitive to the density of states [7] as well as to the height of the surface.

Electron ripples inside a rectangular 2D box can be predicted from $P(x,y)$ from equation (4.60) (setting $n_z = 0$ to recover a two-dimensional case). The values of n_x, n_y appropriate to the ripples in Figure 4.3 are not known, but will be influenced by two facts:

1) First, the energy of the electrons being seen by the STM tip is the Fermi energy of the copper, plus or minus the voltage applied to the tip.
2) Second, the energy scale, suggested by equation (4.60): $E_n = [h^2/8mL^2](n_x^2 + n_y^2)$ (which needs a small modification for unequal sides L_x, L_y) of the trapped electron levels, should be offset below the Fermi energy of copper by 0.44 eV. This is known to be the location of the bottom of the band of the 2D surface electrons on the (111) face of copper.

Electrons in a 3D "Quantum Dot"

Equations (4.59) and (4.60) are applicable to the electron and hole states in semiconductor "quantum dots", which are used in biological research as color-coded fluorescent markers. Typical semiconductors for this application are CdSe and CdTe.

A "hole" (missing electron) in a full energy band behaves very much like an electron, except that it has a positive charge, and tends to float to the top of the band. That is, the energy of the hole increases oppositely to the energy of an electron.

The rules of nanophysics that have been developed so far are also applicable to holes in semiconductors. To create an electron-hole pair in a semiconductor requires an energy at least equal to the energy bandgap, E_g, of the semiconductor.

This application to semiconductor quantum dots requires L in the range of 3–5 nm, the mass m must be interpreted as an effective mass m^*, which may be as small as 0.1 m_e. The electron and hole particles are generated by light of energy

$$hc/\lambda = E_{n,\text{electron}} + E_{n,\text{hole}} + E_g. \tag{4.61}$$

Here the first two terms depend strongly on particle size L, as L^{-2}, which allows the color of the light to be adjusted by adjusting the particle size. The bandgap energy, E_g, is the minimum energy to create an electron and a hole in a pure semiconductor. The electron and hole generated by light in a bulk semiconductor may form a bound state along the lines of the Bohr model, described above, called an exciton. However, as the size of the sample is reduced, the Bohr orbit becomes inappropriate and the states of the particle in the 3D trap, as described here, provide a correct description of the behavior of quantum dots.

4.6.6
2D Bands and Quantum Wires

2D Band

A second physical situation that often arises in modern semiconductor devices is a carrier confined in one dimension, say z, to a thickness d and free in two dimensions, say x and y. This is sometimes called a quantum well.

In this case

$$\psi_n(x,y,z) = (2/d)^{1/2} \sin(n_z \pi z/d) \exp(ik_x x) \exp(ik_y y), \tag{4.62}$$

and the energy of the carrier in the nth band is

$$E_n = (h^2/8md^2) \, n_z^2 + \hbar^2 k_x^2/2m + \hbar^2 k_y^2/2m. \tag{4.63}$$

In this situation, the quantum number n_z is called the sub-band index and for $n = 1$ the carrier is in the first sub-band. We discuss later how a basic change in the electron's motion in a semiconductor band is conveniently described with the introduction of an effective mass m^*.

Quantum Wire

The term Quantum Wire describes a carrier confined in two dimensions, say z and y, to a small dimension d (wire cross section d^2) and free to move along the length of the wire, x. (Qualitatively this situation resembles the situation of a carrier moving along a carbon nanotube, or silicon nanowire, although the details of the bound state wavefunctions are different.)

In the case of a quantum wire of square cross section,

$$\psi_{nn}(x,y,z) = (2/d) \sin(n_y \pi y/d) \sin(n_z \pi z/d) \exp(ik_x x), \tag{4.64}$$

and the energy is

$$E_n = (h^2/8md^2) \, (n_y^2 + n_z^2) + \hbar^2 k_x^2/2m. \tag{4.65}$$

It is possible to grow nanowires of a variety of semiconductors by a laser assisted catalytic process, and an example of nanowires of indium phosphide is shown in Figure 4.5 [8].

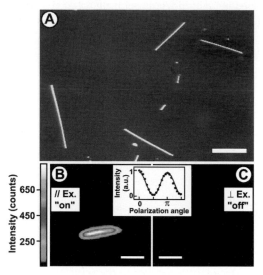

Figure 4.5 Indium phosphide nanowires[8]. InP nanowires grown by laser-assisted catalytic growth, in 10, 20, 30 and 50 nm diameters, were studied by Atomic Force Microscope image (A) and also by observation of photoluminescence (B) and (C) under illumination by light with energy $hc/\lambda > E_g$. The bandgap is about 1.4 eV. In (A), the white scale bar is 5 µm, so the wires are up to 10 µm in length. (B) and (C) gray scale represantation of light emitted from 20 nm diameter InP nanowire excitation in panel (B) with bandgap light linearly polarized along the axis of the wire produces a large photoluminescence, but (C) no light is emitted when excited by bandgap light linearly polarized *perpendicular* to wire axis. Inset shows dependence of emission intensity on polarization angle between light and wire axis.

The InP wires in this experiment are single crystals whose lengths are hundreds to thousands of times their diameters. The diameters are in the nm range, 10–50 nm. The extremely anisotropic shape is shown in Figure 4.2 to lead to extremely polarization-dependent optical absorption.

Since these wires have one long dimension, they do not behave like quantum dots, and the light energy is not shifted from the bandgap energy. It is found that the nanowires can be doped to produce electrical conductivity of N- and P-types, and they can be used to make electron devices.

4.6.7
The Simple Harmonic Oscillator

The simple harmonic oscillator (SHO) represents a mass on a spring, and also the relative motions of the masses of a diatomic molecule. (It turns out that it is also relevant to many other cases, including the oscillations of the electromagnetic field between fixed mirror surfaces.) In the first two cases the mass is m, the spring constant can be taken as k, and the resonant frequency in radians per second is

$$\omega = (k/m)^{1/2}. \tag{4.66}$$

Treated as a nanophysics problem, one needs to solve the Schrodinger equation (4.30) for the corresponding potential energy (to stretch a spring a distance x):

$$U(x) = kx^2/2. \tag{4.67}$$

This is a more typical nanophysics problem than that of the trapped particle, in that the solutions are difficult, but, because of their relevance, have been studied and tabulated by mathematicians.

The solutions to the SHO in nanophysics are:

$$\psi_n = A_n \exp(-m\omega x^2/2\hbar) H_n(x) \tag{4.68}$$

where the $H_n(x)$ are well-studied polynomial functions of x, and

$$E_n = (n + 1/2)\,\hbar\omega, \quad n = 0, 1, 2, 3, \tag{4.69}$$

These wavefunctions give an oscillatory probability distribution $P_n(x) = \psi_n^{*}\psi_n$ that has $n + 1$ peaks, the largest occurring near the classical turning points, $x = \pm\sqrt{2E/k}$. For large n, $P_n(x)$ approaches the classical $P(x)$:

$$P(x)dx = dx/v = \frac{dx}{\sqrt{(2/m)(E-m\omega^2 x^2/2)}}, \tag{4.70}$$

where v is the classical velocity. Since v goes to zero at the turning points, $P(x)$ becomes very large at those points. This feature is approached for large n in the solutions (4.68).

For small n, these functions differ substantially from the classical expectation, especially in giving a range of positions for the mass even in the lowest energy configuration, $n = 0$.

An important point here is that there is an energy $\hbar\omega/2$ (*zero point energy*) for this oscillator even in its lowest energy state, $n = 0$. Also, from equation (4.68), this oscillating mass does not reside solely at $x = 0$ even in its ground state. It has an unavoidable fluctuation in position, given by the Gaussian function $\exp(-m\omega x^2/2\hbar)$, which is close to that predicted by the Uncertainty Principle, equation (4.22).

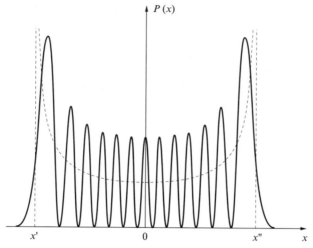

Figure 4.6 Probability $P_n(x)$ density for $n = 13$ state of simple harmonic oscillator. Dashed line is classical $P(x)$, with classical turning points x' and x'' indicated

4.6.8
Schrodinger Equation in Spherical Polar Coordinates

A more substantive change in the appearance of the Schrodinger equation occurs in the case of spherical polar coordinates, which are appropriate to motion of an electron in an atom. In this case, the energy U depends only on the radius r, making the problem spherically symmetric. In standard notation, where $x = r\sin\theta \cos\varphi$, $y = r\sin\theta \sin\varphi$, $z = r\cos\theta$, with θ and φ, respectively, the polar and azimuthal angles:

$$\frac{-\hbar^2}{2m}\frac{1}{r^2}\frac{\partial}{\partial r}\left(r^2\frac{\partial\psi}{\partial r}\right) - \frac{\hbar^2}{2mr^2}\left[\frac{1}{\sin\theta}\frac{\partial}{\partial\theta}\left(\sin\theta\frac{\partial\psi}{\partial\theta}\right) + \frac{1}{\sin^2\theta}\frac{\partial^2\psi}{\partial\varphi^2}\right] + U(r)\psi = E\psi \quad (4.71)$$

4.7
The Hydrogen Atom, One-electron Atoms, Excitons

The Schrodinger equation is applied to the hydrogen atom, and any one-electron atom with nuclear charge Z, by choosing $U = -kZe^2/r$, where k is the Coulomb constant. It is found in such cases of spherical symmetry that the equation separates into three equations in the single variables r, θ, and φ, by setting

$$\psi = R(r)f(\theta)g(\varphi). \quad (4.72)$$

The solutions are conventionally described as the quantum states Ψ_{n,l,m,m_s}, specified by quantum numbers n, l, m, m_s.

The principal quantum number n is associated with the solutions $R_{n,l}(r) = (r/a_o)^l$ $\exp(-r/na_o)\mathscr{L}_{n,l}(r/a_o)$ of the radial equation. Here $\mathscr{L}_{n,l}(r/a_o)$ is a Laguerre polynomial in $\rho = r/a_o$, and the radial function has $n-l-1$ nodes. The parameter a_o is identical to its value in the Bohr model, but it no longer signifies the exact radius of an orbit. The energies of the electron states of the one-electron atom, $E_n = -Z^2 E_o/n^2$ (where $E_o = 13.6\,eV$, and Z is the charge on nucleus) are unchanged from the Bohr model. The energy can still be expressed as $E_n = -kZe^2/2r_n$, where $r_n = n^2 a_o/Z$, and $a_o = 0.0529\,nm$ is the Bohr radius.

The lowest energy wavefunctions $\Psi_{n,l,m,m}$ of the one-electron atom are listed in Table 4.1 [9].

Table 4.1 One-electron wavefunctions in real form [9]

Wavefunction designation	Wavefunction name, real form	Equation for real form of wavefunction*, where $\rho = Zr/a_o$ and $C_1 = Z^{3/2}/\sqrt{\pi}$
Ψ_{100}	1s	$C_1 e^{-\rho}$
Ψ_{200}	2s	$C_2 (2-\rho) e^{-\rho/2}$
$\Psi_{21,\cos\varphi}$	$2p_x$	$C_2\, \rho \sin\theta \cos\varphi\, e^{-\rho/2}$
$\Psi_{21,\sin\varphi}$	$2p_y$	$C_2\, \rho \sin\theta \sin\varphi\, e^{-\rho/2}$
Ψ_{210}	$2p_z$	$C_2\, \rho \cos\theta\, e^{-\rho/2}$
Ψ_{300}	3s	$C_3 (27-18\rho +2\rho^2)\, e^{-\rho/3}$
$\Psi_{31,\cos\varphi}$	$3p_x$	$C_3 (6\rho-\rho^2) \sin\theta \cos\varphi\, e^{-\rho/3}$
$\Psi_{31,\sin\varphi}$	$3p_y$	$C_3(6\rho-\rho^2) \sin\theta \sin\varphi\, e^{-\rho/3}$
Ψ_{310}	$3p_z$	$C_3(6\rho-\rho^2) \cos\theta\, e^{-\rho/3}$
Ψ_{320}	$3d_{z^2}$	$C_4\, \rho^2 (3\cos^2\theta -1)\, e^{-\rho/3}$
$\Psi_{32,\cos\varphi}$	$3d_{xz}$	$C_5\, \rho^2 \sin\theta \cos\theta \cos\varphi\, e^{-\rho/3}$
$\Psi_{32,\sin\varphi}$	$3d_{yz}$	$C_5\, \rho^2 \sin\theta \cos\theta \sin\varphi\, e^{-\rho/3}$
$\Psi_{32,\cos2\varphi}$	$3d_{x^2-y^2}$	$C_6\, \rho^2 \sin^2\theta \cos2\varphi\, e^{-\rho/3}$
$\Psi_{32,\sin2\varphi}$	$3d_{xy}$	$C_6\, \rho^2 \sin^2\theta \sin2\varphi\, e^{-\rho/3}$

* $C_2 = C_1/4\sqrt{2}$, $C_3 = 2C_1/81\sqrt{3}$, $C_4 = C_3/2$, $C_5 = \sqrt{6}C_4$, $C_6 = C_5/2$.

$$\Psi_{100} = \left(Z^{3/2}/\sqrt{\pi}\right) \exp(-Zr/a_o) \tag{4.73}$$

represents the ground state.

The probability of finding the electron at a radius r is given by $P(r) = 4\pi r^2 \Psi^2_{100}$, which is a smooth function easily seen to have a maximum at $r = a_o/Z$. This is not an orbit of radius a_o, but a spherical probability cloud in which the electron's most probable radius from the origin is a_o. There is no angular momentum in this wavefunction!

This represents a great correction in concept, and in numerics, to the Bohr model. Note that Ψ_{100} is real, as opposed to complex, and therefore the electron in this state has no orbital angular momentum. Both of these features correct errors of the Bohr model.

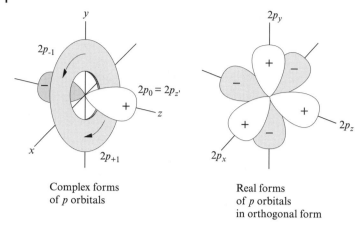

Complex forms
of *p* orbitals

Real forms
of *p* orbitals
in orthogonal form

Figure 4.7 2p wavefunctions in schematic form. Left panel,
complex forms carry angular momentum. Right panel, linear
combinations have the same energy, now assume aspect of
bonds

The $n = 2$ wavefunctions start with Ψ_{200}, which exhibits a node in r, but is spherically symmetric like Ψ_{100}. The first anisotropic wavefunctions are:

$$\Psi_{21,\pm 1} = R(r)f(\theta)g(\varphi) = C_2\,\rho\,\sin\theta\,e^{-\rho/2}\exp(\pm i\varphi), \tag{4.74}$$

where $\rho = Zr/a_o$.

These are the first two wavefunctions to exhibit orbital angular momentum, here $\pm \hbar$ along the *z*-axis. Generally

$$g(\varphi) = \exp(\pm im\varphi), \tag{4.75}$$

where m, known as the magnetic quantum number, represents the projection of the orbital angular momentum vector of the electron along the *z*-direction, in units of $\hbar$. The orbital angular momentum **L** of the electron motion is described by the quantum numbers l and m.

The orbital angular momentum quantum number l has a restricted range of allowed integer values:

$$l = 0,1,2\ldots,n{-}1. \tag{4.76}$$

This rule confirms that the ground state, $n = 1$, has zero angular momentum. In the literature the letters s,p,d,f,g, respectively, are often used to indicate $l = 0,1,2,3$, and 4. So a 2s wavefunction has $n = 2$ and $l = 0$.

The allowed values of the magnetic quantum number m depend upon both n and l according to the scheme

$$m = -l,-l + 1,\ldots,\ (l{-}1),l. \tag{4.77}$$

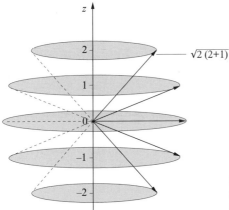

$\sqrt{2\,(2+1)}$

Figure 4.8 Five allowed orientations of angular momentum $l = 2$, length of vector and z-projections in units of $\hbar$. Azimuthal angle is free to take any value

There are $2\,l + 1$ possibilities. For $l = 1$, for example, there three values of m: $-1, 0$, and 1, and this is referred to as a "triplet state". In this situation the angular momentum vector has three distinct orientations with respect to the z-axis: $\theta = 45°$, $90°$ and $135°$. In this common notation, the $n = 2$ state (containing 4 distinct sets of quantum numbers) separates into a "singlet" (2s) and a "triplet" (2p).

For each electron there is also a spin quantum number S with projection

$$m_s = \pm 1/2. \tag{4.78}$$

These strange rules, which are known to accurately describe the behavior of electrons in atoms, enumerate the possible distinct quantum states for a given energy state, n.

Following these rules one can see that the number of distinct quantum states for a given n is $2n^2$. Since the Pauli exclusion principle for electrons (and other Fermi particles) allows only one electron in each distinct quantum state, $2n^2$ is also the number of electrons that can be accommodated in the nth electron shell of an atom. For $n = 3$ this gives 18, which is seen to be twice the number of entries in Table 4.1 for $n = 3$.

A further peculiarity of angular momentum in nanophysics is that the vector **L** has length $L = \sqrt{(l(l + 1))}\;\hbar$ and projection $L_z = m\hbar$. A similar situation occurs for the spin vector **S**, with magnitude $S = \sqrt{(s(s + 1))}\;\hbar$ and projection $m_s\hbar$. For a single electron $m_s = \pm 1/2$. In cases where an electron has both orbital and spin angular momenta (for example, the electron in the $n = 1$ state of the one-electron atom has only S, but no L), these two forms of angular momentum combine as $J = L + S$, which again has a strange rule for its magnitude: $J = \sqrt{(j(j + 1))}\;\hbar$.

The wavefunctions $\Psi_{21,\pm 1} = C_2\,\rho\,\sin\theta\;e^{-\rho/2}\exp(\pm i\varphi)$ are the first two states having angular momentum. A polar plot of $\Psi_{21,\pm 1}$ has a node along z, and looks a bit like a doughnut flat in the xy plane.

The sum and difference of these states are also solutions to Schrodinger's equation, for example

$$\Psi_{211} + \Psi_{21-1} = C_2\,\rho\,\sin\theta\,e^{-\rho/2}\,[\exp(i\varphi) + \exp(-i\varphi)] = C_2\,\rho\,\sin\theta\,e^{-\rho/2}\,2\cos\varphi. \quad (4.79)$$

This is just twice the $2p_x$ wavefunction in Table 4.1. This linear combination is exemplary of all the real wavefunctions in Table 4.1, where linear combinations have canceled the angular momenta to provide a preferred direction for the wavefunction.

A polar plot of the $2p_x$ wavefunction (4.79), shows a node in the z-direction from the $\sin\theta$ and a maximum along the x-direction from the $\cos\varphi$, so it is a bit like a dumb-bell at the origin oriented along the x-axis. Similarly the $2p_y$ resembles a dumb-bell at the origin oriented along the y-axis.

These real wavefunctions, in which the $\exp(im\varphi)$ factors have been combined to form $\sin\varphi$ and $\cos\varphi$, are more suitable for constructing bonds between atoms in molecules or in solids, than are the equally valid (complex) angular momentum wavefunctions. The complex wavefunctions which carry the $\exp(im\varphi)$ factors are essential for describing orbital magnetic moments such as occur in iron and similar atoms. The electrons that carry orbital magnetic moments usually lie in inner shells of their atoms.

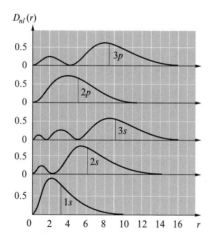

Figure 4.9 Radial wavefunctions of one-electron atoms exhibit n–ℓ–1 nodes, as illustrated in sketch for (bottom to top) 1s, 2s, 3s, 2p and 3p wavefunctions

Magnetic Moments

The magnetic properties of atoms are important in nanotechnology, and lie behind the function of the hard disk magnetic memories, for example. A magnetic moment

$$\mu = i\mathbf{A} \quad (4.80)$$

is generated by a current loop bounding an oriented area **A**. The magnetization **M** is the magnetic moment per unit volume. A magnetic moment produces a dipole magnetic field similar to that of an ordinary bar magnet. The magnetization of a bar of magnetite or of iron is caused by internal alignment of huge numbers of the atomic magnetic moments. These, as we will see, are generated by the orbital motion of electrons in each atom of these materials.

How does an orbiting electron produce a magnetic moment? The basic formula $\mu = iA$ for a magnetic moment can be rewritten as $(dq/dt)\pi r^2 = (ev/2\pi r)\pi r^2 = (e/2m)mrv$. Finally, since mrv is the angular momentum L, we get

$$\mu = (e/2m)\mathbf{L}. \tag{4.81}$$

This is the basic gyro-magnetic relation accounting for atomic orbital magnetism. Recalling that the z-component of $\mathbf{L}$ is $L_z = m\hbar$, we see that the electron orbital motion leads to $\mu_z = (e\hbar/2m_e)\, m$. The Bohr Magneton μ_B is defined as $\mu_B = e\hbar/2m_e = 5.79 \times 10^{-5}$ eV/T. The magnetic moment arising from the electron spin angular momentum is described by a similar formula,

$$\mu_z = g(e\hbar/2m_e)\, m_s. \tag{4.82}$$

Here g, called the gyromagnetic ratio, is $g = 2.002$ for the electron spin and $m_s = \pm 1/2$. The energy of orientation of a magnetic moment μ in a magnetic field B is

$$U = -\mu \cdot \mathbf{B}, \tag{4.83}$$

where the dot product of the vectors introduces the cosine of the angle between them. Thus, the difference in energy between states $m_s -1/2$ and $m_s = -1/2$ for an electron spin moment in a magnetic field B (taken in the z-direction) is $\Delta E = g\mu_B B$.

The reasons for the alignment of atomic magnetic moments in iron and other ferromagnetic substances, which underlies the operation of the magnetic disk memory, will be taken up later in connection with basic symmetry of particles under exchange of positions.

Positronium and Excitons

In the treatment of the hydrogen and one-electron atoms, it has been assumed that the proton or nucleus of charge Z is fixed, or infinitely massive. This simplification is not hard to correct, and a strong example is that of positronium, a bound state of a positron (positive electron) and an electron. These two particles have opposite charges but equal masses, and then jointly orbit around their center of mass, half the distance between the plus and minus charges. It is an exercise to see that the Bohr radius for positronium is $2a_0$ and its binding energy is $E_0/2$. All of the results for the hydrogenic atom can be transcribed to this case by interpreting r as the inter-particle distance and interpreting m to the central mass:

$$1/m_{cm} = 1/m_1 + 1/m_2. \tag{4.84}$$

The classic case is positronium, but a more relevant case is the exciton.

The exciton is the bound state of a photoexcited electron and photoexcited hole in a semiconductor, produced momentarily by light illumination of energy greater than the band gap. The electron and hole particles thus produced orbit around each other, having exactly equal and opposite charges. (In a short time the charges unite, giving off a flash of light.) In the medium of the semiconductor, however, the electron and hole assume effective masses, respectively, m^*_e and m^*_h, which are typically significantly different from the free electron mass.

If, as is typical, one mass, say the electron mass, is small, say 0.1 of the free electron mass, then, from equation (4.84), m_{cm} becomes small. Because the interparticle Bohr radius is inversely dependent on mass m_{cm}, this makes the Bohr radius large, roughly $10\,a_o$. In turn, because the bound state energy is $-kZe^2/2r$, this makes the energy small, roughly $0.1\,E_o$.

4.8
Fermions, Bosons and Occupation Rules

Schrodinger's equation tells us that, in physical situations described by potential energy functions U, there are quantum energy states available for particles to occupy. It turns out that particles in nature divide into two classes, called fermions and bosons, which differ in the way they can occupy quantum states. Fermions follow a rule that only one particle can occupy a fully described quantum state, such as a hydrogen electron state described by n, l, m_l, and m_s. This rule was first recognized by Pauli, and is called the Pauli exclusion principle. In a system with many states and many fermion particles to fill these states, the particles first fill the lowest energy states, increasing in energy until all particles are placed. The highest filled energy is called the Fermi energy E_F.

For bosons, there is no such rule, and any number of particles can fall into exactly the same quantum state. Such a condensation of many photons into a single quantum state is what happens in the operation of a laser.

When a nanophysical system is in equilibrium with a thermal environment at temperature T, then average occupation probabilities for electron states are found to exist. In the case of fermions, the occupation function (the Fermi–Dirac distribution function, f_{FD}), is

$$f_{FD} = \left[\exp\left(\frac{E-E_F}{kT}\right) + 1\right]^{-1}. \tag{4.85}$$

For photons, the corresponding Bose–Einstein distribution function, f_{BE}, is

$$f_{BE} = \left[\exp\left(\frac{hc}{\lambda kT}\right) - 1\right]^{-1}. \tag{4.86}$$

References

[1] N. Bohr, Phil. Mag. **26**, 1 (1913).

[2] E. Schrodinger, Annalen der Physik **79**, 361 (1926).

[3] L. de Broglie, Ann. de phys. **3**, 22 (1925).

[4] C. Davisson and L. Germer, Proc. Natl. Acad. Sci. **14**, 619 (1928).

[5] W. Heisenberg, Zeit. fur. Phys. **43**, 172 (1927).

[6] Courtesy IBM Research, Almaden Research Center. Unauthorized use not permitted.

[7] E. L. Wolf, *Principles of Electron Tunneling Spectroscopy*, (Oxford, New York, 1989), page 319.

[8] Reprinted with permission from Wang, J.S. Gudiksen, X. Duan, Y. Cui, and C.M. Lieber, Science **293**, 1455 (2001). Copyright 2001 AAAS.

[9] F. J. Pilar, Elementary Quantum Chemistry, (Dover, New York., 2001), p. 125.

5
Quantum Consequences for the Macroworld

Chemical matter is made of atoms, which are constructed according to the strange rules of nanophysics. The stability of atoms against radiative collapse was first predicted in the Bohr hydrogen atom model, by the arbitrary imposition of a quantum condition on angular momentum. The Schrodinger equation explains the stability of chemical matter in more detail. An essential additional principle is the Pauli statement that only one electron can occupy a fully described quantum state. This is the "building-up principle" that describes the shell structure of atoms, and the chemical table of the elements.

5.1
Chemical Table of the Elements

The rules governing the one-electron atom wavefunction $\Psi_{n,l,m,m}$ and the Pauli exclusion principle, which states that only one electron can be accommodated in a completely described quantum state, are the basis for the Chemical Table of the Elements. As we have seen, the strange rules of nanophysics allow $2n^2$ distinct states for each value of the principal quantum number, n. There are several notations to describe this situation. The "K shell" of an atom comprises the two electrons of $n = 1$ ($1s^2$), followed by the "L shell" with $n = 2$ ($2s^2 2p^6$); and the "M shell" with $n = 3$ ($3s^2 3p^6 3d^{10}$). These closed shells contain, respectively, 2, 8, and 18 electrons.

In the Chemical Table of Mendeleyev, one notable feature is the stability of filled "electron cores", such as those which occur at $Z = 2$ (He, with a filled K shell), and $Z = 10$ (Ne, with filled K and L shells).

The situation of a single electron beyond a full shell configuration, such as sodium, potassium, rubidium, and cesium, can be roughly modeled as an ns electron moving around the rare gas core described with an effective charge Z', less than Z. That Z' is reduced results from the shielding of the full nuclear charge Z by the inner closed shell electrons. It is remarkable that interactions between electrons in large atoms can be in many cases ignored.

These rules of nanophysics are believed to account for the schematics of the chemical table of the elements, and, as well, to the properties of chemical compounds. It is a logical progression to expect that the larger aggregations of molecules

Nanophysics and Nanotechnology: An Introduction to Modern Concepts in Nanoscience. Edward L. Wolf
Copyright © 2004 WILEY-VCH Verlag GmbH & Co. KGaA, Weinheim
ISBN: 3-527-40407-4

that are characteristic of biology are also understandable in the framework of Schrodinger quantum mechanics.

5.2
Nano-symmetry, Di-atoms, and Ferromagnets

A further and profound nanophysical behavior, with large consequences for the macroworld, is based on the identical nature of the elementary particles, such as electrons.

It is easy to understand that ionic solids like KCl are held together by electrostatic forces, as the outer electron of K fills the outer electron shell of Cl, leading to an electrostatic bond between K^+ and Cl^-.

But what about the diatomic gases of the atmosphere, H_2, O_2, and N_2? The bonding of these symmetric structures, called covalent, is entirely nanophysical in its origin. It is strange but true that the same symmetry-driven electrostatic force that binds these di-atoms is also involved in the spontaneous magnetism of iron, cobalt and other ferromagnetic metals.

5.2.1
Indistinguishable Particles, and their Exchange

The origin of the symmetric covalent bond is the indistinguishable nature of electrons, for which labels are impossible. If two electrons are present in a system, the probability distributions $P(x_1,x_2)$ and $P(x_2,x_1)$ must be identical.

No observable change can occur from exchanging the two electrons. That is,

$$P(x_1,x_2) = P(x_2,x_1) = |\psi_{n,m}(x_1,x_2)|^2, \tag{5.1}$$

from which it follows that either

$$\psi_{n,m}(x_2,x_1) = \psi_{n,m}(x_1,x_2) \quad \text{(symmetric case), or} \tag{5.2}$$

$$\psi_{n,m}(x_2,x_1) = -\psi_{n,m}(x_1,x_2) \quad \text{(antisymmetric case).} \tag{5.3}$$

If we apply this idea to the two non-interacting electrons in, for example, a 1D trap, with the wavefunction (4.39)

$$\psi_{n,m}(x_1,x_2) = A^2 \sin(n\pi x_1/L) \sin(m\pi x_2/L) = \psi_n(x_1) \, \psi_m(x_2),$$

we find that this particular two-particle wavefunction is neither symmetric nor antisymmetric. However, the combinations of symmetric and antisymmetric two-particle wavefunctions

$$\psi_S(1,2) = [\psi_n(x_1) \, \psi_m(x_2) + \psi_n(x_2) \, \psi_m(x_1)]/\sqrt{2}, \text{ and} \tag{5.4}$$

$$\psi_A(1,2) = [\psi_n(x_1)\,\psi_m(x_2) - \psi_n(x_2)\,\psi_m(x_1)]\,/\sqrt{2}, \tag{5.5}$$

respectively, are correctly symmetric and anti-symmetric.

Fermions

The antisymmetric combination ψ_A, equation (5.5), is found to apply to electrons, and to other particles, including protons and neutrons, which are called *fermions*. By looking at ψ_A in the case $m = n$, one finds $\psi_A = 0$.

The wavefunction for two fermions in exactly the same state, is zero! This is a statement of the Pauli exclusion principle: only one Fermi particle can occupy a completely specified quantum state.

Bosons

For other particles, notably photons of electromagnetic radiation, the symmetric combination $\psi_S(1,2)$, equation (5.4), is found to occur in nature. Macroscopically large numbers of photons can have exactly the same quantum state, and this is important in the functioning of lasers. Photons, alpha particles, and helium atoms are examples of Bose particles, or *bosons*.

The Bose condensation of rare gas atoms has been an important research area in the past several years. The superfluid state of liquid helium, and the superconducting ground state of electron pairs, can also be regarded as representing Bose condensates, in which macroscopically large numbers of particles have exactly the same quantum numbers.

Electron pairs and the Josephson effect are available only at low temperatures, but they enable a whole new technology of computation, in which heat generation (energy consumption) is almost zero. So the question, is a refrigerator worth its expense, is becoming more of a realistic economic choice, as energy density in the silicon technology continues to rise with Moore's Law.

Orbital and Spin Components of Wavefunctions

To return to covalent bonds of di-atoms, it is necessary to complete the description of the electron state, by adding its spin projection $m_s = \pm 1/2$. It is useful to separate the space part $\phi(x)$ and the spin part χ of the wavefunction, as

$$\psi = \phi(x)\chi. \tag{5.6}$$

For a single electron $\chi = \uparrow$ (for $m_s = 1/2$) or $\chi = \downarrow$ (for $m_s = -1/2$). For two electrons there are two categories, $S = 1$ (parallel spins) or $S = 0$ (antiparallel spins). While the $S = 0$ case allows only $m_s = 0$, the $S = 1$ case has three possibilities, $m_s = 1, 0, -1$, which are therefore referred to as constituting a "spin triplet".

A good notation for the spin state is $\chi_{S,m}$, so that the spin triplet states are

$$\chi_{1,1} = \uparrow_1\uparrow_2, \quad \chi_{1,-1} = \downarrow_1\downarrow_2, \quad \text{and} \quad \chi_{1,0} = 1/\sqrt{2}\,(\uparrow_1\downarrow_2 + \downarrow_1\uparrow_2) \quad \text{(spin triplet)} \tag{5.7}$$

For the singlet spin state, $S = 0$, one has

$$\chi_{0,0} = 1/\sqrt{2}\,(\uparrow_1\downarrow_2 - \downarrow_1\uparrow_2) \quad \text{(spin singlet)} \tag{5.8}$$

Inspection of these makes clear that the spin triplet ($S = 1$) is symmetric on exchange, and the spin singlet ($S = 0$) is antisymmetric on exchange of the two electrons.

Since the *complete* wavefunction (for a fermion like an electron) must be antisymmetric for exchange of the two electrons, this can be achieved in two separate ways:

$$\psi_A(1,2) = \phi_{sym}(1,2)\,\chi_{anti}(1,2) = \phi_{sym}(1,2)\ \ S = 0\ \text{(spin singlet)} \tag{5.9}$$

$$\psi_A(1,2) = \phi_{anti}(1,2)\,\chi_{sym}(1,2) = \phi_{anti}(1,2)\ \ S = 1\ \text{(spin triplet).} \tag{5.10}$$

The structure of the orbital or space wavefunctions $\phi_{sym,anti}(1,2)$ here is identical to those shown above for $\psi_{S,A}(1,2)$, equations (5.4) and (5.5).

5.2.2
The Hydrogen Molecule, Di-hydrogen: The Covalent Bond

Consider two protons, (labeled a, and b, assume they are massive and fixed) a distance R apart, with two electrons. If R is large and we can neglect interaction between the two atoms, then, following the discussion of Tanner [1],

$$[-(\hbar^2/2m)\,(\nabla_1^2 + \nabla_1^2) + U(r_1) + U(r_2)]\psi = (\mathcal{H}_1 + \mathcal{H}_2)\,\psi = E\psi \tag{5.11}$$

where the rs represent the space coordinates of electrons 1 and 2. Solutions to this problem, with no interactions, can be $\psi = \psi_a(x_1)\psi_b(x_2)$ or $\psi = \psi_a(x_2)\psi_b(x_1)$ (with $\psi_{a,b}$ the wavefunction centered at proton a,b) and the energy in either case is $E = E_a + E_b$.

The interaction between the two atoms is the main focus, of course. The interactions are basically of two types. First, the repulsive interaction $ke^2/r_{1,2}$, with $r_{1,2}$ the spacing between the two electrons. Secondly, the attractive interactions of each electron with the 'second' proton, proportional to $(1/r_{a,2} + 1/r_{b,1})$. The latter attractive interactions, primarily occurring when the electron is in the region between the two protons, and can derive binding from both nuclear sites at once, stabilize the hydrogen molecule (but destabilize a ferromagnet).

Altogether we can write the interatom interaction as

$$\mathcal{H}_{int} = ke^2[1/R + 1/r_{1,2} - 1/r_{a,2} - 1/r_{b,1}]. \tag{5.12}$$

To get the expectation value of the interaction energy the integration, below, following equation (4.38) must extend over all six relevant position variables q. (Spin variables are not acted on by the interaction.)

$$<E_{int}> = \int \psi^* \mathscr{H}_{int} \, \psi \mathrm{d}q. \tag{5.13}$$

The appropriate wavefunctions have to be overall antisymmetric. Thus, following equations (5.9) and (5.10), the symmetric $\phi_{sym}(1,2)$ orbital for the antisymmetric $S = 0$ (singlet) spin state, and the anti-symmetric $\phi_{anti}(1,2)$ orbital for the symmetric $S = 1$ (triplet) spin state.

The interaction energies are

$$<E_{int}> = A^2 \left(K_{1,2} + J_{1,2} \right) \quad \text{for} \quad S = 0 \text{ (spin singlet)} \tag{5.14}$$

$$<E_{int}> = B^2 \left(K_{1,2} - J_{1,2} \right) \quad \text{for} \quad S = 1 \text{ (spin triplet), where} \tag{5.15}$$

$$K_{1,2}(R) = \iint \phi_a^*(x_1) \phi_b^*(x_2) \mathscr{H}_{int} \, \phi_b(x_2) \phi_a(x_1) \mathrm{d}^3 x_1 \mathrm{d}^3 x_2 \tag{5.16}$$

$$J_{1,2}(R) = \iint \phi_a^*(x_1) \phi_b^*(x_2) \mathscr{H}_{int} \, \phi_a(x_2) \phi_b(x_1) \mathrm{d}^3 x_1 \mathrm{d}^3 x_2 \tag{5.17}$$

The physical system will choose, for each spacing R, the state providing the most negative of the two interaction energies. For the hydrogen molecule, the exchange integral $J_{1,2}$ is negative, so that the covalent bonding occurs when the spins are anti-parallel, in the spin singlet case.

Covalent bonding and covalent anti-bonding, purely nanophysical effects
The qualitative understanding of the result is that in the spin singlet case the orbital wavefunction is symmetric, allowing more electron charge to locate halfway between the protons, where their electrostatic energy is most favorable. A big change in electrostatic energy (about $2J_{1,2}$) is linked (through the exchange symmetry requirement) to the relative orientation of the magnetic moments of the two electrons. This effect can be summarized in what is called the exchange interaction

$$\mathscr{H}_e = -2J_e \, \mathbf{S}_1 \cdot \mathbf{S}_2 \tag{5.18}$$

In the case of the hydrogen molecule, J is negative, giving a negative *bonding* interaction for *antiparallel* spins. The parallel spin configuration is repulsive or *antibonding*. The difference in energy between the bonding and antbonding states is about 9 eV for the hydrogen molecule at its equilibrium spacing, $R = 0.074$ nm. The bonding energy is about 4.5 eV.

This is a huge effect, ostensibly a magnetic effect, but actually a combination of fundamental symmetry and electrostatics. The covalent bond is what makes much of matter stick together.

The covalent bond as described here is a short-range effect, because it is controlled by the overlap of the exponentially decaying wavefunctions from each nucleus. This is true even though the underlying Coulomb force is a long-range effect, proportional to $1/r^2$.

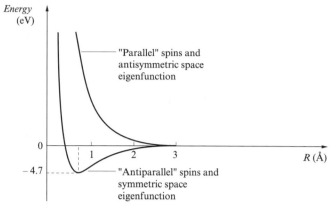

Figure 5.1 Energy curves for bonding and anti-bonding states of the hydrogen molecule. The bonding state requires anti-parallel spins. The equilibrium separation is 0.074 nm

Covalent bonds exist not only between two s-states ($\ell = 0$), as in hydrogen, but also between two p-states ($\ell = 1$) as well as between an s and a p state. Organic molecules are, on the whole, bonded together in a covalent fashion.

Ferromagnetism, a purely nanophysical effect
In other cases J is positive, making the energy a favorable arrangement one of *parallel spins*. This leads to *ferromagnetism*, a cooperative state of matter in which huge numbers of magnetic moments are all locked parallel, leading to a macroscopic magnetization, M. It is important to recognize that the driving force for the spin alignment is electrostatic, the exchange interaction, represented by equation (5.18).

5.3
More Purely Nanophysical Forces: van der Waals, Casimir, and Hydrogen Bonding

We have just seen that the nanophysical exchange interaction leads both to the strong electrostatic covalent bond and to ferromagnetism. These are completely non-classical effects, depending on the wave-particle nature of matter and the symmetry of the wavefunction against exchange of particles. An ionic bond, such as that which stabilizes NaCl, table salt, is a more transparent, classical Coulomb force effect. The ionic bond is characteristic of many solids that dissolve in water. The covalent bond dominates the internal bonding of most of the molecules of organic chemistry, which includes the molecules of living things. To be sure, bonds are often partially ionic and partially covalent. For example, the basically covalent H_2O molecule has a partial negative charge on O, and thus has a permanent electric dipole moment. Covalent and ionic bonds are strong, giving several electron volts of binding energy. For example, the 6 eV bandgap energy of diamond is the energy to remove one electron from (create a hole in) an $n = 2$ tetrahedral sp^3 bond, a process

in which the electron does not leave the solid but becomes free to move in the conduction band of the solid.

We now consider further bonding forces that occur in the macro-world that stem
from its nanophysical basis. These are the polar and van der Waals (dispersion)
forces, the Casimir force, and hydrogen bonding. These are all weaker forces than
the covalent and ionic bonding forces. The Casimir force in particular is vanishingly
small except for extremely closely spaced surfaces; yet it is that regime that nanotechnology may exploit in the future. However, for two metal surfaces spaced by
10 nm, there is an attractive pressure of about 1 atmosphere, 100 kPa, from the
Casimir effect.

The polar, van der Waals, and hydrogen bonding forces are important in allowing
molecules to bond together into larger, if less strongly bonded, assemblies such as
proteins, and they play large roles in biology.

5.3.1
The Polar and van der Waals Fluctuation Forces

Many molecules, for example water, H_2O, are partially polar and have a net electric
dipole moment, $\mathbf{p} = q\mathbf{a}$. Here q is the electric charge and $\mathbf{a}$ is its displacement vector.

The vector electric field $\mathbf{E}_d$ produced by an electric dipole $\mathbf{p}$ can be written

$$\mathbf{E}_d = k\left[\frac{\mathbf{p}}{r^3} - \frac{3(\mathbf{p} \cdot \mathbf{r})}{r^5}r\right]. \tag{5.19}$$

This dipolar field $\mathbf{E}_d$ resembles the magnetic $\mathbf{B}$ field of a bar magnet. Here k is the
Coulomb constant $k = (4\pi\varepsilon_o)^{-1}$, and the overall strength of the dipole field is kp/r^3.

A second dipole $\mathbf{p}_2$ will interact with this field as $U = -\mathbf{p}_2 \cdot \mathbf{E}_d$. So the interaction
energy U is about

$$U \sim -kp_1p_2/r^3 \sim -ke^2a_o^2/r^3, \tag{5.20}$$

taking a value $p = ea_o$.

The force associated with this interaction is $-dU/dr$; this dipole-dipole force is
attractive and the strength is

$$F \sim -3\,kp_1p_2/r^4. \tag{5.21}$$

For this reason polar molecules, for example, water molecules, will attract each
other and may condense to a liquid or solid, giving an attractive interaction of about
0.42 eV per molecule (the heat of vaporization of liquid water). (Since this cohesive
energy involves one molecule interacting with its group of near neighbors, the pair-
wise interaction at liquid water density is around 0.1 eV.) Another effect of importance is that polar molecules will cluster around an ion (an ion–dipole interaction),
creating a sheath which decreases the strength of the Coulomb field of the ion at
large distances. This is a screening effect. Hydration sheaths around ions in water

are well known, and for example, will affect the rate at which an ion will diffuse in a liquid.

Electric polarizability of neutral atoms and molecules
An atom in an electric field develops an induced dipole moment

$$\mathbf{p} = \alpha \mathbf{E}, \tag{5.22}$$

where α is the electric polarizability. The simplest case is the hydrogen atom, for which

$$\alpha = (4.5)\, a_o^3/k, \tag{5.23}$$

where k, the Coulomb constant, appears because of the S.I. Units. The polarizability α increases with the number of electrons, in a complicated fashion. An estimate of the effect by London [2, 3] gives

$$\alpha \sim 2ne^2 z^2{}_{av}/I_1. \tag{5.24}$$

Here n is the number of electrons, $z^2{}_{av}$ is the average of the square of the position of the electron along the applied electric field, and I_1 is the first (lowest) ionization energy.

The important point is that the polarizability α becomes very large as the number of electrons on the atom increases. (This is probably why the IBM group chose xenon ($Z = 54$) as the atom to move along the surface of (111) Cu in their early STM nanofabrication work.)

This effect then gives an interaction energy between a dipolar atom and a non-polar atom, because the field of the dipolar atom will induce a moment in the non-polar atom. This interaction energy varies as $U \sim -\alpha k^2 p_1^2/r^6$.

The effects described so far come from static dipoles inherent in the bonding of molecules such as water.

Dipolar fluctuations of neutral and symmetric atoms
The purely quantum-mechanical van der Waals effect, in contrast, is a dynamic interaction, resulting from the fact that the electron in an atom is a charged particle in chaotic motion (not in a regular orbit). As such it represents a fluctuating electric dipole of strength $p = ea_o$. All of the electrons in an atom or molecule take part in this effect, not just those in the bonding, and the result is a fluctuating dipole electric field in the region around the atom or molecule, described as above.

The van der Waals effect comes from the *induced* dipoles in *neighboring* matter, arising from the fluctuating dipolar field and the polarizability of all its electrons. In this way fluctuations on one atom become correlated with fluctuations on neighboring atoms, and an attractive interaction U results. Calculations of the interaction energy U for two hydrogen atoms $U_{vdw} = -6.5\, ke^2 a_o^5/r^6$. (This result can be approached from the interaction between two hydrogen atoms considered above in connection with covalent bonding, by expanding $1/R$ in powers of x, y, and z.) (There is a correction to this energy estimate relating to the transit time for light between the two interacting electrons, treated by Casimir and Polder, which can be neglected.)

For two many-electron atoms the result is

$$U_{vdw} = -1.5 \, k^2 \frac{\alpha_A \alpha_B}{r^6} \frac{I_A I_B}{I_A + I_B}, \tag{5.25}$$

where I_A and I_B are the first ionization energies and the α_A and α_B are the polarizabilities. It is noteworthy that these polarizabilities are proportional to the numbers n of electrons in each atom, so this energy can be hundreds of times larger than the van der Waals interaction between two hydrogen atoms. This interaction is important in the liquefaction of symmetric atoms, such as rare gases, and in molecular solids, as shown in Figure 5.2 [after 4].

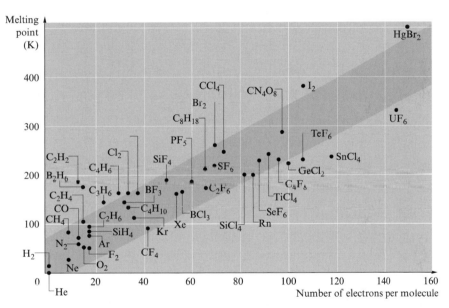

Figure 5.2 Melting points of molecular solids versus number of electrons per molecule. The melting point is a measure of the cohesive energy, presumably here largely originating in van der Waals interactions [after 4]

The van der Waals effect is a short-range interaction, varying as r^{-6}, while the Coulomb energy varies as r^{-1}. However, if a single molecule or atom is interacting with a large body such as a plane or large solid sphere, the dependence of the interaction on the spacing Z from the point to plane or sphere, evident after integrating the interaction energy over the locations in the extended body, will become $1/Z^3$. A nice summary of dipolar and van der Waals forces in chemical and biological contexts is given by Rietman [5].

A number of extended geometries have been analyzed based on the r^{-6} interaction. Figure 5.3 shows a few of these schematically. For more information see Rietman [5].

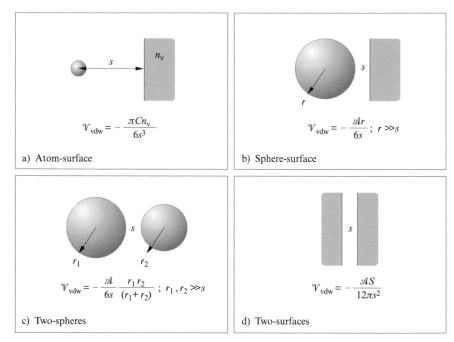

a) Atom-surface

$$\mathcal{V}_{vdw} = -\frac{\pi C n_v}{6 s^3}$$

b) Sphere-surface

$$\mathcal{V}_{vdw} = -\frac{\mathcal{A} r}{6 s} \; ; \; r \gg s$$

c) Two-spheres

$$\mathcal{V}_{vdw} = -\frac{\mathcal{A}}{6 s} \frac{r_1 r_2}{(r_1 + r_2)} \; ; \; r_1, r_2 \gg s$$

d) Two-surfaces

$$\mathcal{V}_{vdw} = -\frac{\mathcal{A} S}{12 \pi s^2}$$

Figure 5.3 van der Waals interaction energies for several extended geometries. In these panels n_v represents the number of atoms per unit volume, $\mathcal{A}$ is the Hamaker constant, a tabulated material parameter, and S represents the area of two facing surfaces [after 6]

The van der Waals interaction between extended bodies typically shows a slower inverse distance dependence than the underlying atom–atom interaction. The Hamaker constant $\mathcal{A}$, a tabulated material parameter useful in estimating interactions, is typically in the range of 0.2 eV – 3.12 eV, and increases with the number of atoms per unit volume in the material in question. For more details the reader is referred to Drexler [6].

5.3.2
The Casimir Force

The Casimir force [7] operates between metallic surfaces, forcing them together. It is an electromagnetic effect which has to do with the modes of oscillation of the electromagnetic field in an enclosed region.

Two parallel mirrors spaced a distance Z apart will allow standing electromagnetic waves to build up, propagating in the z-direction perpendicular to the surfaces, when $Z = n\lambda/2$. If we think of the transverse electric field $E(z,t)$, then E must be zero at each mirror surface, $z = 0$ and $z = Z$. Modes like this are what produce the external light beam in a laser pointer. (One of the mirrors is partially metallized and lets some light propagate out.) This resonance is similar to a standing wave on a violin string, which we can think of in terms of simple harmonic motion. For the violin string in its fundamental mode, the transverse motion of the string mass is a simple harmonic motion, in which the energy is fully kinetic when the string is at zero displacement, and fully potential when the string is fully displaced. For both the electromagnetic modes and the resonant frequencies on the violin string, the shorter the length (the mirror spacing Z or the length L of the violin string) the fewer modes that can be supported, because the longest wavelength available in either case is twice the cavity or string length.

Equations (4.68) and (4.69) indicate that the harmonic oscillator in nanophysics acquires two strange characteristics: a zero point motion and a zero point energy, features which are consistent with the uncertainty principle. *The lowest energy state of the oscillator is not zero, but $\hbar\omega/2$.*

It turns out that the electromagnetic modes in the cavity between the two mirrors act like nanophysical oscillators, and have for this reason a zero point energy and also fluctuations in electric and magnetic fields even at temperature zero. From this point of view, even an empty cavity between mirrors has an energy U, namely: $\hbar\omega/2$, summed over all the modes which are allowed in the cavity. If the cavity width is Z, the only available frequencies are $n(c/2L)$, where $n = 1,2,3,\ldots$ so that all frequencies less than $c/2L$ are unavailable. Equivalently, wavelengths greater than $2L$ are excluded, and as L is reduced, more wavelengths, and more zero point energies, are excluded.

The energy cost U for the electromagnetic fluctuations in the cavity is reduced as L is reduced, and this leads to a force $F = -dU/dZ$ to collapse the cavity. This scenario has been experimentally verified by recent accurate observations of the Casimir force.

A careful calculation of the attractive Casimir force between parallel mirrors [7,8] gives

$$F_C = -\frac{\pi^2 \hbar c}{240} \frac{1}{Z^4} \tag{5.26}$$

Here c is the speed of light and Z is the spacing of the mirrors.

This represents an attractive force per unit area (negative pressure) which is very small at large spacings Z, but rises to a value of about 1 atmosphere, 100 kPa, for $Z = 10$ nm. 10 nm is a spacing that might well arise in the nanomachines of the future, so the Casimir force may well be of engineering relevance.

An easier geometry to arrange reliably is a sphere of radius R spaced a distance z from a plane surface. In this case, the Casimir force on a sphere is [8]

$$F_{Cs} = -\frac{\pi^3 \hbar c}{360} R \frac{1}{z^3}. \tag{5.27}$$

This equation was confirmed in a silicon nanotechnology experiment sketched (not to scale) in Figure 5.4. A stage on which two polysilicon capacitor plates are delicately suspended to rotate an angle θ by two silicon torsion fibers, is moved upward (using piezoelectric stage, below) toward 100 µm radius sphere R, reducing the spacing z to as small as 75 nm. Plate spacing of the capacitors is 2 µm, established by etching away a grown 2 µm SiO_2 layer on which the two polysilicon upper plates had been deposited.

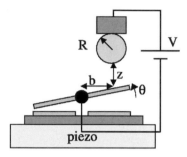

Figure 5.4 Schematic of measurement of the Casimir force between sphere and planar surface, measured with $V = 0$ [8]. Planar capacitor plates on paddle rotate against torsion fiber (center black dot) whose restoring force is calibrated using V and known Coulomb force between sphere and plane. Rotation angle θ is measured by capacitance difference between paddle plates and lower fixed plates, whose spacing t is 2 µm. Piezo stage lifts paddle assembly toward sphere, adjusting z down to 75 nm

The micromachined balanced capacitor assembly, grown on a single crystal silicon chip, is shown in Figure 5.5 [8]. The upper capacitor plate is a 3.5 µm thick polysilicon rectangle deposited onto 2 µm thick silicon dioxide grown on a crystal immediately below. After deposition of the upper plate, the silicon dioxide is etched out, leaving the upper plate freely suspended.

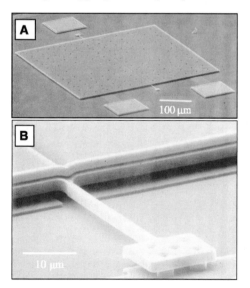

Figure 5.5 Micromachined planar capacitor plates freely suspended to rotate above silicon crystal on twin silicon torsion fibers [8]. (A) Rectangular paddle 500μ on a side is freely suspended 2 µm above Si chip by twin Si fibers, front and back. (B) Detail of paddle (upper rotating capacitor plate) showing its separation from Si chip and showing the front Si suspending torsion fiber. 200μ radius metallized sphere (not shown) is placed a variable distance z above center of right portion of the paddle (see Figure 5.4)

The force between the sphere and the right portion of the paddle is determined by the twist angle θ of the fiber, which in turn is determined from the difference in capacitances to the separate lower capacitor plates. (The capacitance of a parallel plate capacitor is $\varepsilon_o A/t$, where A is the area, t the plate spacing and ε_o the permittivity of vacuum, 8.854×10^{-12} Farads/meter.) The torsion constant of the fiber was calibrated by the twist angles θ observed with voltage biases V applied to the sphere, and with z large enough that the Casimir force is negligible. The Coulomb force between a sphere of radius R and a plane at spacing z is a known function of R, V and z, to which the measurements were carefully fitted, thus calibrating the force vs. twist angle θ.

Measured forces vs. sphere-plane spacing z are shown in Figure 5.6 [after 8].

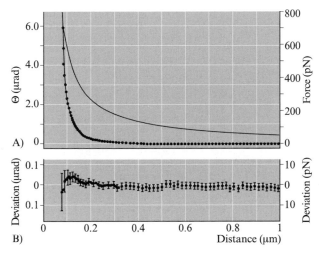

Figure 5.6 Measured Casimir force between sphere and plane vs. spacing z, shown by points fitted to theoretical curve [after 8]. (A) Data points and Casimir theory curve (lower trace) vs. spacing z. Upper curve is force vs. z for $V = 136\,mV$, where Coulomb force matches Casimir force at closest approach (76 nm). (B) Deviation between measured points and fitted Casimir theory curve, on expanded scale

Two important corrections to the theory equation (5.27) have been made in achieving the fit shown in Figure 5.6. The first is that the gold metal evaporated onto the surfaces is not in fact a perfect mirror, and allows electromagnetic waves of very high frequency to pass through. This effect was corrected for, by making use of the tabulated optical constants for gold. The second effect is that the gold mirror surfaces are not perfectly flat but have a roughness amplitude of 30 nm as measured with an atomic force microscope. These two corrections are carefully made and explained in [8], leaving no doubt that the Casimir force has been accurately measured and is an important effect for metal surfaces whose spacing is on the 100 nm scale. This scale is relevant to nanotechnology.

5.3.3
The Hydrogen Bond

The hydrogen bond occurs in situations where the slightly positive H in the polar water molecule locates midway between negative charges, which may occur on the negative end of other polar molecules. The sharing of this partially charged H atom with two negative entities is a bit like the sharing of a negative electron in a covalent bond between two protons. Hydrogen bonding occurs frequently when polar water molecules are important, and indeed the structure of ice is described as hydrogen bonded.

Hydrogen bonding is weak, which is appropriate to its role in joining the twin strands of double helix DNA. This weakness allows the helical strands to be separated easily in models of DNA replication (see Figure 5.7) involving the DNA polymerase engine, similar to the RNA polymerase engine whose force measurement was depicted in Figure 3.12.

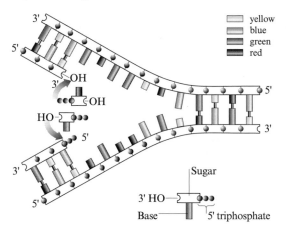

Figure 5.7 Models of DNA replication fork, showing breaking of hydrogen bonds between upper and lower bases (center of figure) as DNAP engine (left) pulls double helical DNA from right to left [after 9]. Hydrogen bonds are depicted here as smaller diameter cylinders connecting the larger cylinders representing the four different bases A, C, G, and T, which, however, bond only as complementary pairs AT and CG. The AT pairs form double hydrogen bonds, while the CG pairs form triple hydrogen bonds

In Figure 5.7 [after 9] the "rails" depict the strong repeating nucleotide units that form the single DNA strands. The repeating unit consists of a deoxyribose cyclic sugar molecule, an ionic phosphate $(PO_4)^{2-}$, and the "base". The bases are Adenine, Cytosine, Guanine and Thymine, A, C, G, and T, of which only complementary combinations AT and CG hydrogen bond from one strand to the other. In Figure 5.7 the groups of three small spheres represent triphosphates. Adenosine triphosphate

(ATP) is well known as a source of energy, which is stored in the negatively charged phosphate bonds, and is released when these bonds are broken.

Following [9], nucleoside triphosphates serve as a substrate for DNA polymerase, according to the mechanism shown on the top strand. Each nucleoside triphosphate is made up of three phosphates (represented here by small spheres), a deoxyribose sugar (rectangle) and one of four bases (larger cylinders of differing lengths). The three phosphates are joined to each other by high-energy bonds, and the cleavage of these bonds during the polymerization reaction releases the free energy needed to drive the incorporation of each nucleotide into the growing DNA chain.

Hydrogen bonds are important in many other areas of large-molecule chemistry and biology. A good source of information on these topics is Ball [10].

5.4
Metals as Boxes of Free Electrons: Fermi Level, DOS, Dimensionality

To a first approximation, a metal or a semiconductor can be regarded as a 3D "box" containing the free electrons released from the outermost orbits of the atoms. Atoms such as Na or Ca give up one or two of their valence electrons when they form a solid. These freed electrons form a delocalized electron gas in the "box". To a remarkable degree the electrons do not interfere with each other, and the dominant effects come from the Pauli principle which means that each of the electrons go into a separate energy state, up to the Fermi energy.

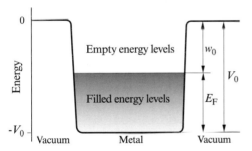

Figure 5.8 Metal as a 3D box filled with non-interacting electrons up to the Fermi energy E_F, following the Pauli exclusion principle. The total depth of the potential well is V_o, the sum of E_F and the work function, here labeled w_o

Such a box is sketched in Figure 5.8, surrounded by the potential barrier described by the work function, described as w or ϕ, which contains the electrons inside the solid. The empty box is an unexpectedly good representation for the electron motion, it turns out, because of the *periodic spatial variation* of the realistic potential, $U(x,y,z)$, coming from the array of atoms in the real solid. The *periodic* arrangement of the charged ions very much reduces their disruption of electron motions, as we will discuss later.

The work function is usually several electron volts, and certainly contains the electrons in the metal. The states of the 3D trap considered earlier, are a good starting point.

$$\psi = (2/L)^{3/2} \sin(n_1\pi x/L) \sin(n_2\pi y/L) \sin(n_3\pi z/L). \tag{5.28}$$

Since $\sin kx = (e^{ikx} - e^{-ikx})/i2$, we can consider these states to be superpositions of oppositely directed traveling waves $\psi_+ = \exp(ikx)$ and $\psi_- = \exp(-ikx)$. This is of course the same as in the case of standing waves on a violin string. Here the moving waves $\psi_{+,-}$ are more fundamental for a description of conduction processes.

The singly most important aspect of nanophysics for this situation is the Pauli exclusion principle, which states that only one electron can occupy a fully described state. This means that if we add a large number of electrons to the box, the quantum numbers will be come very large, and also the energy of the successively filled states will be large. The energy is given by equation 4.60:

$$E_n = [h^2/8mL^2](n_x^2 + n_y^2 + n_z^2). \tag{5.29}$$

It is convenient to rewrite this equation as

$$E_n = E_o \, r^2, \tag{5.30}$$

as an aid to counting the number of states filled out to an energy E, in connection with Figure 5.9. In coordinates labeled by integers n_x, n_y, and n_z, constant energy surfaces are spherical and exactly two electron states occupy a unit volume. Since the states are labeled by positive integers, only one octant of a sphere is involved. The number N of states out to radius r is

$$N = (2)\,(1/8)\,(4\pi r^3/3) = \pi r^3/3 = (\pi/3)\,(E/E_o)^{3/2}, \tag{5.31}$$

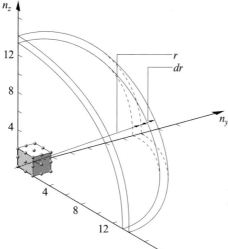

Figure 5.9 Constant energy surfaces in quantum number space for particle in a 3D potential well, a starting representation for electrons in a metal. There are two electron states per lattice point. The number of states N out to radius r is therefore $(2/8)(4\pi/3)r^3$, where $r = E/E_o$, and $E_o = h^2/8mL^2$ (see eq. 5.29)

where $E_o = h^2/8mL^2$, for a box of side L. This is equivalent to

$$E_F = (h^2/8m)\,(3N/\pi\,L^3)^{2/3} = (h^2/8m)\,(3N/\pi V)^{2/3}. \qquad (5.32)$$

Setting $N/V = n$, the number of free electrons/m^3, it is straightforward to see that

$$dN/dE = g(E) = (3n/2)\,E^{1/2}E_F^{-3/2} \qquad (5.33)$$

is the density of electron states per unit volume at energy E.

The Fermi velocity u_F is defined by $mu_F^2/2 = E_F$, and the Fermi temperature is defined by $kT_F = E_F$. The Fermi velocity and the Fermi temperature are much larger than their thermal counterparts, both important consequences of the Pauli principle in filling the electron states in a box representing the metal.

At zero temperature, states below E_F are filled and states above E_F are empty. At nonzero temperatures the occupation is given by the Fermi function $f = [\exp(\{E-E_F\}/kT) + 1]^{-1}$. The width in energy of transition of f from 1 to 0 is about kT.

Some of these features are sketched in Figure 5.10.

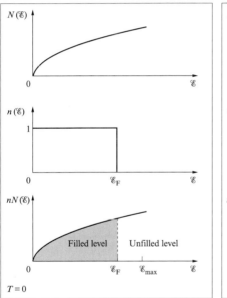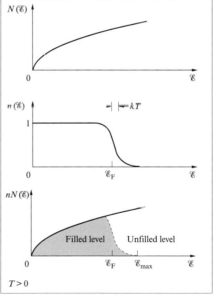

Figure 5.10 Density of states $g(E)$ and occupation $f(E)$ at $T = 0$ (left) and T nonzero (right) in the 3D case

A collection of data on Fermi energies and Fermi temperatures is given in Table 5.1.

Table 5.1 Fermi energy E_F, Fermi temperature T_F, and free electron density $n = N/V$ for metals

Element	N/V ($\times 10^{28}$ m^{-3})	Fermi energy (eV)	Fermi temperature ($\times 10^4$ K)
Al	18.1	11.7	13.6
Ag	5.86	5.53	6.41
Au	5.90	5.55	6.43
Cu	8.47	7.06	8.19
Fe	17.0	11.2	13.0
K	1.40	2.13	2.47
Li	4.70	4.77	5.53
Mg	8.61	7.14	8.28
Mn	16.5	11.0	12.8
Na	2.65	3.26	3.78
Sn	14.8	10.3	11.9
Zn	13.2	9.50	11.0

We have considered the 3D case in the material above. Dimensionality makes a difference in the behavior.

In the 1D case, as a quantum wire, N atoms in a length L, the density of states is

$$g(E) = (L/\pi) \, (m/2\hbar^2 E)^{1/2}. \tag{5.34}$$

In the 2D case, the density of states is a constant, independent of energy.

$$g(E) = \text{constant} \tag{5.35}$$

5.5
Periodic Structures (e.g. Si, GaAs, InSb, Cu): Kronig–Penney Model for Electron Bands and Gaps

Electrical conduction in metals and semiconductors is a topic of considerable importance in modern electronic technology, which is based on semiconductors such as silicon and gallium arsenide. An advanced understanding of the behavior of conduction electrons is available in these important cases. The "band theory" of solids is based on an assumption that electron behavior can be treated as a one-electron problem of motion in an average potential U which is primarily determined by the array of partially ionized atoms aligned on the crystalline lattice of the solid. Detailed predictions of physical properties are available for crystalline solid materials of interest, including many semiconductors and metals. It has even proven possible to predict the transition temperatures, T_c of superconducting metals.

This understanding of the behavior of electrons in such solids is based on concepts of nanophysics extended to include the periodic nature of the potential U that acts upon an electron in a crystalline solid. In large part, the basic formulas for current density $J = nev$, drift velocity $v = (e\tau/m)E$, mobility $\mu = e\tau/m$, resistivity $(1/\rho = ne^2\tau/m)$ etc. are retained, but the band theory allows improvement in their accuracy by providing an "effective mass" m^* which replaces the electron mass m_e, as well as a relative dielectric constant which multiplies ε_0. The band theory also provides a new basis for calculating the mean free path and scattering time τ in pure crystalline solids.

A basic description of a free electron of specified kinetic energy and momentum in one dimension is $\Psi_x = (1/L)^{1/2}\exp[i(kx-\omega t)]$, where $k = p/\hbar = 2\pi/\lambda$ and $E = \hbar\omega$. Here Ψ_x is normalized to provide unit probability to find a free electron in a length L. Ψ_x is a solution of the 1D Schrodinger equation with $U = 0$. How can such a solution be modified to describe an electron in a solid? Surely the correct wavefunction will have some periodic property to match the periodic potential.

It is easy to show that wavefunctions for such a periodic potential are of the form

$$\psi = u_k(x)\exp(ikx),\tag{5.36}$$

where $u_k(x)$ exhibits the periodicity of the atomic lattice, a, and where the allowed values of k are $n\pi/L$, where $n = 1,\ldots,N$. This wavefunction can still represent a particle of speed $\hbar k/m$.

The answer is that if $U(x)$ is a periodic function based on the atomic spacing of the crystalline solid, the free electron description still applies to certain ranges of energy E, while in other ranges of energy, called "energy gaps", no electron states are allowed.

Assume a row of $N = L/a$ atoms on the x-axis. Then $U(x + na) = U(x)$, with $n = 1,\ldots,L/a$.

With the periodic potential $U(x)$, new phenomena of "allowed energy bands" and "forbidden energy gaps" appear in the dependence $E(k)$ of electron energy on wavevector k. These effects are mathematically similar to the existence of *pass bands* in the impedance characteristics of periodic transmission lines. These new effects are best viewed in comparison to $E = \hbar^2 k^2/2m = n^2 h^2/8mL^2$ for the case $U = 0$, as in the 1D infinite square well.

An introduction to this new basic behavior is afforded by a simple soluble model, due to Kronig and Penney[11]. This model assumes a linear array of N atoms spaced by a along the x-axis, $0 < x < Na = L$. The 1D potential $U(x)$ is of a square-wave form, see Figure 5.11, with period a. We assume the origin of the x-axis is at the center of one atom;

$$U(x) = -V_0/2, \quad 0 < x < (a-w)/2$$
$$V_0/2, \quad 0 < x < (a+w)/2.\tag{5.37}$$

represented in this $U(x)$ by symmetric potential wells of depth $-V_0/2$ and width $a-w$, separated by barriers of height $V_0/2$ and width w. Note that this extended periodic

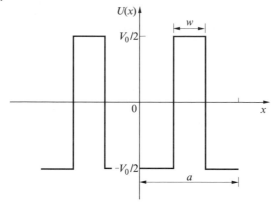

Figure 5.11 A simplified model potential, assumed extended periodically

$U(x)$ is symmetric about $x = 0$, i.e. $U(x) = U(-x)$, and that the average value of U is 0 as in the free particle case.

The solutions (5.36) are compatible with the 1D Schrodinger equation

$$(-\hbar^2/2m)\mathrm{d}^2\Psi/\mathrm{d}x^2 + [U(x) - E]\,\Psi = 0$$

containing the periodic $U(x)$ of (5.37), *only* if the following condition is satisfied:

$$\cos ka = \beta(\sin qa/qa) + \cos qa = R(E), \tag{5.38}$$

where $\beta = V_0 wma/\hbar^2$ and $q = (2mE)^{1/2}/\hbar$. (The simplified form of (5.38) is actually obtained in a limiting process where the potential barriers are simultaneously made higher and narrower, preserving the value of β. This can be described as N δ-functions of strength β.) The parameter β is a dimensionless measure of the strength of the periodic variation.

One easily sees that in (5.38) in the limit $\beta = 0$, the condition $\cos ka = \cos \alpha a$ leads to $k = (2mE)^{1/2}/\hbar$, which recovers the free electron result, $E = \hbar^2 k^2/2m$. Next, if we let β become extremely large, the only way the term $\beta(\sin qa/qa)$ can remain finite, as the equation requires, is for $\sin qa$ to become zero. This requires $qa = n\pi$, or $a(2mE)^{1/2}/\hbar = n\pi$, which leads to $E = n^2 h^2/8ma^2$. These are the levels for a 1D square well of width a (recall that in the limiting process the barrier width w goes to zero, so that each atom will occupy a potential well of width a).

The new interesting effects of band formation occur for finite values of β. Figure 5.12 gives a sketch of the right-hand side, $R(E)$, of (5.38), vs. qa. Solutions of this equation are only possible when $R(E)$ is between -1 and $+1$, the range of the $\cos ka$ term on the left. Solutions for E, limited to these regions, correspond to allowed energy bands (shown unshaded in sketch of Figure 5.12). Note that allowed solutions are possible for $-1 < \cos ka < 1$, which corresponds to $-\pi < ka < \pi$. More generally,

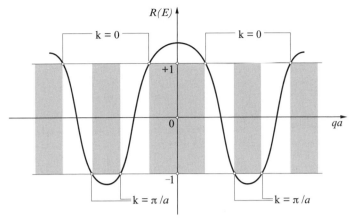

Figure 5.12 Kronig–Penney model, a schematic plot of ordinate $R(E)$ vs. abscissa qa. Allowed solutions (unshaded) occur only when ordinate $R(E)$ has magnitude unity or less

boundaries of the allowed bands are at $k = (+/-)n\pi/a$, $n = 1,2,3\ldots$. A sketch of the allowed $E(k)$ is shown dotted in Figure 5.13.

Bragg scattering of electron waves The special role of $k = \pi/a$ in these equations represents a fundamental physical process, Bragg scattering of the electron wave. When a traveling wave $\exp(ikx)$ is scattered by the atoms located at $x = na$, a coherent reflected wave, $\exp(-ikx)$, can be generated, leading to a standing wave.

The condition for coherent reflection from succeeding atom positions $x = na$ is the 1D analog of the Davisson–Germer experiment mentioned in Chapter 2. The path difference between a wave back-scattered at $x = 0$ and one back-scattered at $x = a$, is $2a$. Thus, for coherent reflection $2a$ must be an integer number of electron wavelengths. This Bragg condition, $n\lambda = 2a$, since $\lambda = 2\pi/k$, is equivalent to

$$k = n\pi/a \tag{5.39}$$

Any wave $\exp(in\pi x/a)$ thus generates a reflected wave $\exp(-in\pi x/a)$, and this means that the only waves present at $k = n\pi/a$ are the linear combinations $\exp(in\pi x/a) \pm \exp(-in\pi x/a)$:

$$\psi(k = n\pi/a) = \cos(n\pi x/a) \text{ or } \sin(n\pi x/a). \tag{5.40}$$

These are *standing waves*, corresponding, respectively to probability densities $P(x)$ peaking at atom positions ($x = na$), or midway between. The $\cos^2(n\pi x/a)$ solution will thus have a lower energy than the $\sin^2(n\pi x/a)$ solution (at the same k). The difference in these energies exactly provides the energy gap between the lower band and the next band.

A second result of Bragg reflection is that, at $k = n\pi/a$, the standing waves give no net particle transport (equal flows to right and left), so the group velocity, $\hbar^{-1}\partial E/\partial k = 0$. The energy bands thus have zero slope at the "zone boundaries", $k = \pm(\pi/a)$.

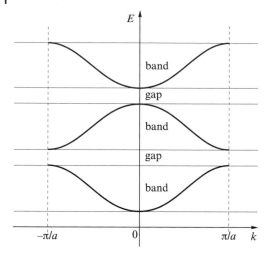

Figure 5.13 Schematic of bands E vs. k in a periodic potential, based on Kronig–Penney model. The bands are restricted in k to values less than π/a. Energy gaps occurring at $k = \pm(\pi/a)$ are also physically understood on the basis of Bragg reflections at $k = \pm(\pi/a)$. Physical arguments easily show that each band accommodates exactly $N/2$ electrons, so that one electron per atom gives a half-filled band and a metal, while two electrons per atom gives a filled band, and an insulator

Returning to discussion of the Kronig–Penney model, it is easy to show that the slope of $E(k)$ is zero at the boundaries, i.e., the group velocity $\partial\omega/\partial k$ goes to zero at the edges of the bands. To show this, expand $\cos ka$ of (5.38) about the points $k = n\pi/a$; and then form the differential dE/dk. Let $ka = n\pi + x$, where x is small; so

$$\cos(ka) = \cos(n\pi + x) = \cos(n\pi)\cos(x) - \sin(n\pi)\sin(x) \approx \cos(n\pi)(1 - x^2/2) = R(E) \quad (5.41)$$

Taking the differentials of last two terms, $-\cos(n\pi)x\,dx = (dR/dE)dE$. Since $dx = a\,dk$, we have, near the points $k = n\pi/a$, the value $dE/dk \approx -a\cos(-\pi)[dR/dE]^{-1}x$. Thus, in the limit $x = 0$, (i.e., at the points $k = n\pi/a$) we have $dE/dk = \hbar\partial\omega/\partial k = 0$. This is actually always the case near a band edge, independent of our simplified model of the periodic $U(x)$.

As shown in the solid curves in Figure 5.13, it is conventional to shift the higher band segments back toward the origin by appropriate multiples of $2\pi/a$, assembling them in the "first zone" $-\pi/a < k < \pi/a$. The solid lines in Figure 5.13 represent the three lowest energy bands. It is clear that the width of the allowed bands decreases as the strength parameter β increases, and that the gaps disappear with β tending to zero.

For N sites the theory predicts $2N$ delocalized electron states, with the factor of 2 coming from the fact that N electrons of spin quantum number $m_s = 1/2$ and N electrons of spin quantum number $m_s = -1/2$ are allowed. The basis for this idea of "filling" the bands is the Pauli principle for fermion particles such as electrons, which states that only one fermion can occupy a single fully specified quantum state. For example, a hydrogen atom state specified by n,l,m,m_s can have only one electron. The electrons of course occupy lowest energy states preferentially. For this reason, N univalent atoms (such as Na) which donate only one electron, will lead to a half-filled band. Such a band gives metallic conduction, because empty $E(k)$ states are present immediately above filled states, allowing for acceleration of the electrons. Conventional metals Na, Cu, Ag, Au etc. have half-filled conduction bands. In these cases, the number of free electrons can be represented as νN, where ν is the valence. On the other hand, the theory predicts that a filled band will lead to no conduction, i.e., an insulator, because no empty states are available to allow motion of the electrons in response to an electric field. Such a situation is present in the pure semiconductors Si, GaAs, and Ge.

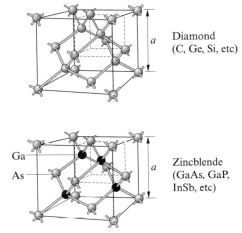

Diamond
(C, Ge, Si, etc)

Ga
As

Zincblende
(GaAs, GaP,
InSb, etc)

Figure 5.14 Diamond and zincblende crystal structures. Each atom is covalently bonded to four nearest neighbors in tetrahedral directions. The directed bonds are linear combinations of s and p orbitals (see Table 4.1), and analogous to directed orbitals sketched in Figure 4.7. Specifically, 2s and $2p^3$ for diamond (as in CH_4) and 3s and $3p^3$ for Si. There are four valence electrons per atom, leaving a band structure with filled bands

In these materials, four outer electrons (two each from 3s and 3p orbitals in Si and GaAs, and two each from 4s and 4p in Ge), are stabilized into four tetrahedral covalent bond orbitals which point from each atom to its four nearest neighbor atoms, located at apices of a tetrahedron.

5.6
Electron Bands and Conduction in Semiconductors and Insulators

The four electrons per atom (eight electrons per cell) which fill the covalent bonds of the diamond-like structure, completely fill the lowest two "valence" bands (because of the spin degeneracy, mentioned above), leaving the third band empty. In concept this would correspond to the first two bands in Figure 5.13 as being completely filled, and thus supporting no electrical conduction. Referring to Figure 5.13, one sees that at $k = 0$, just above the second band, there is a forbidden gap, E_g. There are no states allowing conduction until the bottom of band 3, which is about 1 eV higher in these materials. For this reason, at least at low temperatures, pure samples of these materials do not conduct electricity.

Electrical conductivity at low temperature and room temperature in these materials is accomplished by "doping"; substitution for the four-valent Si or Ge atoms either acceptor atoms of valence 3 or donor atoms of valence 5. In the case of five-valent donor atoms like P, As, or Sb, four electrons are incorporated into tetrahedral bonding and the extra electron becomes a free electron at the bottom of the next empty band. In useful cases, the number of free electrons, n, is just the number of donor atoms, N_D. This is termed an N-type semiconductor. In the case of boron, aluminum, and other three-valent dopant atoms, one of the tetrahedral bonding states is unfilled, creating a "hole". A hole acts like a positive charge carrier, it easily moves as an electron from an adjacent filled bond jumps into the vacant position. Electrical conductivity by holes is dominant in a "P-type semiconductor".

The band structures for Si and GaAs are sketched in Figure 5.15. These results are calculated from approximations to Schrodinger's equation using more realistic 3D forms for U. The curves shown have been verified over a period of years by various experiments.

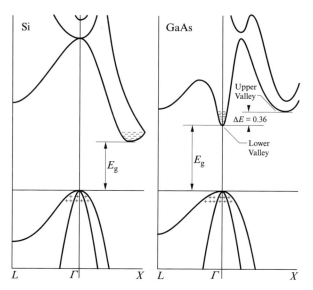

Figure 5.15 Energy band structures for Silicon (left) and GaAs (right). Energy is shown vertically, and k horizontally. The horizontal line marks the top of the filled "valence" bands; in pure samples the upper bands are empty except for thermal excitations (indicated by ++ and −−symbols.) The zero of momentum is indicated as "Γ", and separate sketches are given for E vs k in (111) left and (100) right directions.

In the case of these semiconductors, the charge carriers of importance are either electrons at a minimum in a nearly empty conduction band, or holes at the top of a nearly filled valence band. In either case, the mobility $\mu = e\tau/m^*$, such that $\mathbf{v} = \mu\mathbf{E}$, is an important performance parameter. A high mobility is desirable as increasing the frequency response of a device such as a transistor. A useful quantity, which can be accurately predicted from the band theory, is the effective mass, m^*. This parameter is related to the inverse of the curvature of the energy band. The curvature, $d^2 E/dk^2$ can be calculated, and the formula, simply related to $E = \hbar^2 k^2/2m^*$, is

$$m^* = \hbar^2/d^2 E/dk^2. \tag{5.42}$$

In looking at the energy bands for Si and GaAs in Figure 5.15, one can see that generally the curvature is higher in the conduction band than in the valence band, meaning that the effective mass is smaller and the mobility therefore higher for electrons than for holes. Secondly, comparing Si and GaAs, the curvature in the conduction band minimum is higher in the latter case, leading to a higher mobility for GaAs electrons than in Si. A further aspect is that the conduction band minimum in Si is shifted from $k = 0$, which has important effects especially on the absorption of photons by Si, and similar semiconductors having an "indirect" bandgap. Parameters describing several important semiconductors are collected in Table 5.2.

Table 5.2 Energy gaps and other electronic parameters of important semiconductors

Semiconductor	Band Gap (eV)		Mobility at 300 °K (cm²/volt s)		Effective Mass m^*/m_o		Dielectric Constant ε
	300 °K	0 °K	Electrons	Holes	Electrons	Holes	
C	5.47	5.51	1800	1600	0.2	0.25	5.5
Ge	0.66	0.75	3900	1900	$m_l^* = 1.6$	$m_{lh}^* = 0.04$	16
					$m_t^* = 0.082$	$m_{hh}^* = 0.3$	
Si	1.12	1.16	1500	600	$m_l^* = 0.97$	$m_{lh}^* = 0.16$	11.8
					$m_t^* = 0.19$	$m_{hh}^* = 0.5$	
Grey Tin		~0.08					
AlSb	1.63	1.75	200	420	0.3	0.4	11
GaN	3.5						
GaSb	0.67	0.80	4000	1400	0.047	0.5	15
GaAs	1.43	1.52	8500	400	0.068	0.5	10.9
GaP	2.24	2.40	110	75	0.5	0.5	10
InSb	0.16	0.26	78 000	750	0.013	0.6	17
InAs	0.33	0.46	33 000	460	0.02	0.41	14.5
InP	1.29	1.34	4600	150	0.07	0.4	14

The energy gaps of semiconductors range from about 0.1 to about 5 eV, as indicated in Table 5.2. For devices that operate at room temperature a gap of at least 1 eV is needed to keep the number of thermally excited carriers sufficiently low.

The understanding of semiconductors represented in the sections above is sufficient to reconsider the case of the tunneling diode mentioned in Chapter 1. In more detail, the current voltage characteristic of a tunnel diode in forward bias is shown in Figure 5.16. The forward bias curve for a conventional diode is sketched in a dashed curve in Figure 5.16. At the peak of the anomalous current, marked b in the solid curve of Figure 5.16, electrons in the conduction band on the N-type side are able to tunnel directly into the filled hole states (empty electron states) at the top of the valence band on the P-side. This is possible only if the depletion layer width W is sufficiently small, on the order of 5 nm, and in turn this is possible only if the doping concentrations of donors on the n-side and acceptors on the p-side are sufficiently high. The planar junction between an N-type region and a P-type region in a semiconductor such as Ge contains a "depletion region" separating conductive regions filled with free electrons on the N-side and free holes on the P-side. The width W of the depletion region is

$$W = [2\varepsilon\varepsilon_o V_B (N_D + N_A)/e(N_D N_A)]^{1/2}. \tag{5.43}$$

Here $\varepsilon\varepsilon_o$ is the dielectric constant, e the electron charge, V_B is the energy shift in the bands across the junction, and N_D and N_A, respectively, are the concentrations of donor and acceptor impurities.

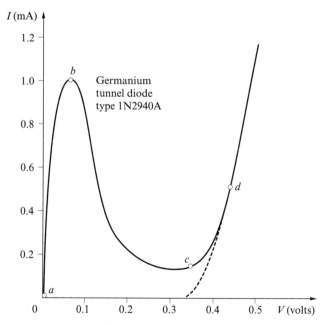

Figure 5.16 Current voltage measurement of germanium tunnel diode (Esaki diode), in forward bias, emphasizing anomalous current peak and negative resistance region (b to c). For an interpretation, see Figure 5.17

Referring to Figure 5.17, the band configurations are illustrated corresponding to bias voltages labeled as a, b, c, and d in Figures 5.16 and 5.17. Positive or forward bias corresponds to raising the energy bands on the n-side relative to those on the p-side, so that electron current flows from right to left.

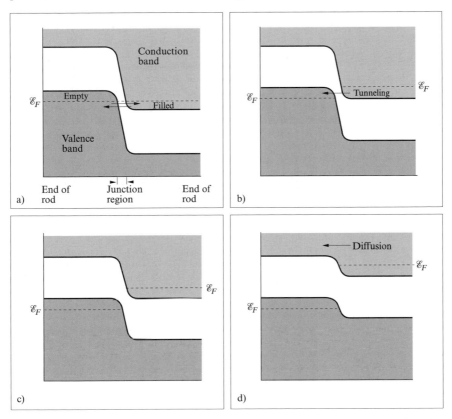

Figure 5.17 Band shifts in forward bias of heavily doped pn junction, showing tunneling as the origin of the anomalous current peak in the Esaki tunnel diode (see Figure 5.16). Panel (b) corresponds to forward bias in which electrons can tunnel directly into filled hole states on the left. At larger forward bias in panel (c) no current can flow because electrons face the energy gap of the p-type region, which does not allow current flow. Finally, in panel (d) the thermal activation of electrons over the barrier leads to the usual exponential growth of current with forward positive bias V

A final topic relates to the concept of resistivity, in the quantum picture. The electron states that we have been dealing with are perfectly conducting, in the sense that the electron wave of such a state maintains a velocity $v = \hbar k/m$. A perfectly periodic potential gives a perfect conductor. Indeed, very pure samples of GaAs, especially epitaxial films, measured at low temperatures, give mobility values of millions, in units of $cm^2/volt \cdot s$. The meaning of the lifetime, τ, in the expression for the mobility, is the lifetime of an electron in a particular k state. The cause of limited-state lifetime τ in pure metals and semiconductors at room temperature is loss of perfect periodicity as a consequence of thermal vibrations of the atoms on their lattice positions. Calculations of this effect in metals, for example, lead to the observed linear dependence of the resistivity ρ on the absolute temperature T.

5.7
Hydrogenic Donors and Acceptors

Pure semiconductors have filled valence bands and empty conduction bands and thus have small electrical conductivity, depending on the size of the energy gap. As mentioned above, larger electrical conductivity is accomplished by "doping"; substitution for the four-valent Si or Ge atoms either acceptor atoms of valence 3 or donor atoms of valence 5.

In the case of five-valent *donor* atoms like P, As, or Sb, four electrons are incorporated into tetrahedral bonding and the fifth, extra, electron, which cannot be accommodated in the already-filled valence band, must occupy a state at the bottom of the next empty band, which can be close to the donor ion, in terms of its position. This free electron is attracted to the donor impurity site by the Coulomb force. The same physics that was described for the Bohr model should apply! However, in the semiconductor medium, the Coulomb force is reduced by the relative dielectric constant, ε. Referring to Table 5.1, values of ε are large, 11.8 and 16, respectively, for Si and Ge.

A second important consideration for the motion of the electron around the donor ion is the *effective mass* that it acquires because of the band curvature in the semiconductor conduction band. These are again large corrections, m^*/m is about 0.2 for Si and about 0.1 for Ge. So the Bohr model predictions must be *scaled* by the change in dielectric constant and also by the change of electron mass to m^*.

The energy and Bohr radius are, from equations (4.3), $E_n = -kZe^2/2r_n$, and $r_n = n^2 a_o/Z$, where $a_o = \hbar^2/mke^2 = 0.053$ nm. Consider the radius first, and notice that its equation contains both k, the Coulomb constant (which will be proportional to $1/\varepsilon$) and the mass. So the corrected Bohr radius will be

$$a_o{}^* = a_o(\varepsilon m/m^*). \tag{5.44}$$

Similarly considering the energy, $E_n = -kZe^2/2r_n = -E_o Z^2/n^2$, $n = 1,2,\ldots$, where $E_o = mk^2 e^4/2\hbar^2$, it is evident that E scales as $m^*/(m\,\varepsilon^2)$:

$$E_n{}^* = E_n[m^*/(m\,\varepsilon^2)]. \tag{5.45}$$

For the case of donors in Si, we find $a^* = 59a_o = 3.13$ nm, and $E_o{}^* = 0.0014\,E_o = 0.0195$ eV. The large scaled Bohr radius is an indication that the continuum approximation is reasonable, and the small binding energy means that most of the electrons coming from donors in Si at room temperature escape the impurity site and are free electrons in the conduction band. An entirely analogous situation occurs with the holes circling acceptor sites. So for the case of the heavily-doped pn junction (Esaki diode), shown in Figure 5.16, the doping values N_D and N_A are essentially equal to numbers of free electrons and holes, as the analysis assumed.

5.8
More about Ferromagnetism, the Nanophysical Basis of Disk Memory

In a ferromagnet, each atom has an electronic magnetic moment, and these moments are all aligned in the same direction. The alignment comes from the electrostatic exchange interaction, when J is positive, discussed above:

$$\mathcal{H}_e = -2J_e\, \mathbf{S}_1 \cdot \mathbf{S}_2. \tag{5.18}$$

The consequence of the spin alignment, forced by the electrostatic interaction, itself related to the symmetry of the wavefunction against exchange, is that the sample has a bulk magnetization, magnetic moment M per unit volume. It turns out that regions of alignment, called *domains*, appear, and the domain sizes adjust as the *domain walls* shift. If the walls are not easily moved, then the "coercive field" H_c is large, see Figure 5.18.

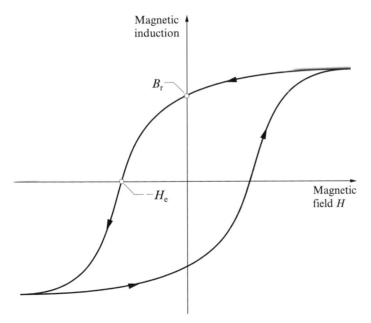

Figure 5.18 Bulk moment M vs. applied magnetic field H, for typical ferromagnet. Hysteretic behavior arises from motion of domain walls. A large remanent M at $H = 0$ is typical of a permanent magnet

In zero applied field, although there is no preferred direction for the domain magnetization, the domains tend to remain oriented, leading to a "remanent magnetization". The limit of large applied field gives the saturation magnetization, when all domains are aligned parallel. As a result of the, sometimes complex, domain structure, the magnetization M of a sample has a characteristic dependence upon applied magnetic field, as sketched in Figure 5.18.

The critical temperature T_c for ferromagnetism is approximately related to the strength of the positive exchange interaction, J_E, by the following expression:

$$J_E = \frac{3k_B T_C}{2zS(S+1)}, \tag{5.46}$$

where z is the number of nearest neighbors, S is the effective spin quantum number of the group of electrons in the atom that constitute its magnetic moment, and k_B is Boltzmann's constant.

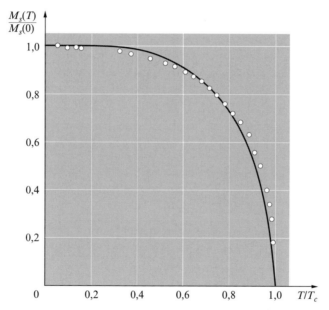

Figure 5.19 The saturation magnetic moment $M_S(T)$ of a typical ferromagnet vs. temperature, normalized to critical temperature T_c. Circles are data for ferromagnet nickel. Solid line comes from solution to equation (5.48)

In thinking about this equation, it is important to realize that the driving force for the effect is the positive exchange interaction J_E, and that the ferromagnetic transition T_C is the consequence.

An understanding of the temperature dependence of the magnetization below T_C can be obtained by making an assumption that the exchange interaction can be simulated by a fictitious internal exchange field B_E. We assume that the saturation magnetization $M_S(T)$ (which occurs within domains) is proportional to this "magnetic field":

$$M_S(T) = B_E/\lambda, \tag{5.47}$$

where λ is a proportionality constant. A self-consistency equation can be obtained, which is solved to determine $M_S(T)$. The magnetization $M_S(T)$ at low temperatures for N spins (say $S = 1/2$) in a magnetic field (say B_E) is $M_S(T) = N\mu_B \tanh(\mu_B B_E/k_B T)$. This is the Brillouin function, which is an accurate description of *paramagnetism*, and μ_B is the Bohr magneton. Making use of (5.44) we get the equation to be solved to find *ferromagnetism*:

$$M_S(T) = N\mu_B \tanh[\mu_B \lambda M_S(T)/k_B T] . \tag{5.48}$$

Solution of this equation for M gives the solid line as plotted in Figure 5.19. This equation reasonably well balances the strength of the exchange interaction, represented by λ, and the thermal agitation represented by $k_B T$, for a *bulk* sample.

An appropriate question for nanotechnology is: how small a piece of iron, or other magnetic material, will still exhibit ferromagnetism, which is a cooperative bulk effect?

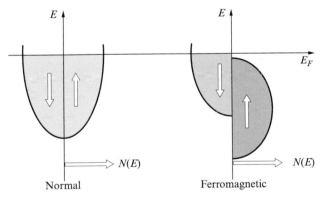

Figure 5.20 Schematic band fillings for normal metal (left) and ferromagnet (right). Normal metal has equal numbers of spin-up and spin-down electrons, here depicted in a partially filled s-band. On the right, a ferromagnet is depicted with relatively narrow d-bands, shifted by $2\mu B_E$, where B_E is the *internal exchange field*, representing the action of the exchange interaction J_E. A portion of the spin-down band is empty, leaving a net excess of spin-up electrons

As a ferromagnetic particle is reduced in size, one can expect it eventually to contain only a single domain. At the "superparamagnetic" limit, it is found that the direction of the magnetic moment becomes unstable and changes its direction over time, depending upon the temperature [12]. Modeling of small ferromagnets has been reported, in connection with a prediction of a maximum magnetic disk data density of $36Gb/in^2$ [13].

Observations [14] in a transmission electron microscope of the magnetization of 10 nm magnetite Fe_3O_4 particles confirmed motion of the magnetization, but on a much slower time scale than the 5ns predicted. It is reported that the superparamagnetic limiting size for $SmCo_5$ is 2nm [14].

In Chapter 6 we will discuss *magnetotactic* bacteria, which have been found to have arrays of 100 nm scale ferromagnets within their single cells. The magnets are near the superparamagnetic limit, and it may be true that the linear array stabilizes even superparamagnetic magnetite particles into magnetic alignment. These bacteria, and the question of the limiting size of magnetic bits written on computer magnetic hard drives, are examples of the nanophysical phenomena of ferromagnetism appearing near the smallest limiting size.

Properties of some common ferromagnets are summarized in Table 5.3 [15].

Table 5.3 Properties of some common ferromagnets. The column "$n_B (0 K)$" is expressed in multiples of the Bohr magneton, $\mu_B = 9.27 \times 10^{-24}$ J/T [15].

Substance	Magnetization M_g, in gauss		$n_B(0 K)$,	Curie temperature,
	Room temperature	0 K	per formula unit	in K
Fe	1707	1740	2.22	1043
Co	1400	1446	1.72	1388
Ni	485	510	0.606	627
Gd	–	2060	7.63	292
Dy	–	2920	10.2	88
MnAs	670	870	3.4	318
MnBi	620	680	3.52	630
MnSb	710	–	3.5	587
CrO_2	515	–	2.03	386
$MnOFe_2O_3$	410	–	5.0	573
$FeOFe_2O_3$	480	–	4.1	858
$NiOFe_2O_3$	270	–	2.4	858
$CuOFe_2O_3$	135	–	1.3	728
$MgOFe_2O_3$	110	–	1.1	713
Eu	–	1920	6.8	69
$Y_3Fe_5O_{12}$	130	200	5.0	560

5.9
Surfaces are different, Schottky barrier thickness $W = [2\varepsilon\varepsilon_o V_B/eN_D]^{1/2}$

Figure 5.14 shows the unit cell of a diamond structure, which occurs in Si and other semiconductors. The structure features complete accommodation of the valence electrons of each atom in directed bonds.

If such a crystal is split in half, exposing two surfaces, the bonding will be disrupted. If the electron bonding structure locally does not change, then there will be "dangling bonds", which will be reactive and unstable. Passivation of such a surface is sometimes carried out by reacting it with hydrogen, to fill the dangling bonds. In other cases the surface may reconstruct into a form which makes use of all of the valence electrons in a modified bonding scheme. The details of what happens will depend upon in which direction the crystal was split. The properties of a (111) surface, as was illustrated in connection with Cu, will be different from those of (110) or (100) surfaces.

A frequent and important surface effect is an *electrical barrier layer* (Schottky barrier). A simple way of thinking of bending of the energy bands at the surface is that the position of the Fermi level relative to the bands may be different at the surface than it is in the interior.

For example, in N-type silicon the Fermi level in the bulk of the material is just below the conduction band edge. At the surface, however, the Fermi level may lie in the middle of bandgap. The surface typically will have a large density of states in the midgap energy range, and electrons from the adjoining N-type material will fall into these states, making the surface electrically negatively charged and leaving a positively charged depletion layer. The equilibrium situation that results is that the conduction band edge at the surface is raised by $eV_B \sim E_g/2$, leaving a depletion region between the surface and the bulk. The width of the Schottky barrier, a depletion region, is

$$W = [2\varepsilon\varepsilon_0 V_B/eN_D]^{1/2}. \tag{5.49}$$

For donor concentrations N_D convenient for bulk device operation, the resulting W is so large that electrical current will not cross the Schottky barrier. So there is an inherent difficulty in making conductive Ohmic contacts to Silicon. One way to correct this difficulty is to make the surface layer heavily doped (N^{++}), so that, according to (5.49), the width W is small enough that electrons can readily tunnel through the barrier.

References

[1] B. K. Tanner, *Introduction to the Physics of Electrons in Solids* (Cambridge, Cambridge, 1996), p. 186.

[2] F. London, Zeit. für Phys. **63**, 245 (1930).

[3] L. Pauling and E. B. Wilson, *Introduction to Quantum Mechanics with Applications to Chemistry* (Dover, New York, 1985) p. 387.

[4] K. S. Krane, *Modern Physics* (2nd Edition) (Wiley, New York, 1995). Figure 11.12, p. 344.

[5] E. A. Reitman, *Molecular Engineering of Nanosystems* (Springer, New York, 2001) p. 71.

[6] K. E. Drexler, *Nanosystems*, (Wiley, New York, 1992) pp. 64–65.

[7] H. G. B. Casimir, Proc. Kon. Ned. Akad. **51**, 793 (1948).

[8] Reprinted with permission from H. B. Chan, V. A. Aksyuk, R. N. Kleiman, D. J. Bishop, and F. Capasso, Science **291**, 1941– 1944 (2001). Published online 8 February 2001 (10.1126/science.1057984). Copyright 2001 AAAS.

[9] B. Alberts, Nature **421**, 431 (2003).

[10] P. Ball, *Designing the Molecular World: Chemistry at the Frontier* (Princeton, Princeton, 1994), p. 151.

[11] R. L. de Kronig and W. G. Penney, Proc. Roy. Soc. Ser. A **130** 499 (1931).

[12] E. F. Kneller and F. E. Luborsky, J. Appl. Phys. **34**, 656 (1963).

[13] S. H. Charap, P. L. Lu, Y. He, IEEE Trans. Magn. **33**, 978 (1997).

[14] S. A. Majetich and Y. Jin, Science **284**, 470 (1999).

[15] C. Kittel, *Introduction to Solid State Physics* (Sixth Edition) (Wiley, New York, 1986). Table 2, p. 429.

6
Self-assembled Nanostructures in Nature and Industry

Atoms and molecules are formed spontaneously when the proper ingredients (particles or atoms) and the proper conditions, such as temperature and pressure, are available.

Atoms on the whole have formed naturally. The nuclei of all atoms, especially those of many protons and neutrons, are believed to have formed in the interior of large stars, under conditions of extreme temperature and pressure. It is believed that the heavy atoms in our solar system have condensed from remains of exploded stars, called supernovas. The statement of Carl Sagan: "We are made of star-stuff" (in terms of the starting atoms) is well documented in decades of work of astronomers and astrophysicists. Self-assembly is definitely the preferred route to the formation of atoms.

Molecules, also, are mostly self-assembled and available in nature. (Certainly this statement is overwhelmingly true on a weight basis; consider the mass of the solar system, but there are now many key molecules that man has invented and produced even to the point of altering the ecosystem, e.g., DDT).

The entire array of biological molecules, from the point of view of science, has remarkably self-assembled. The task of the scientist has been to figure out how it happened! The learning has been rapid, especially after the genetic code represented by DNA was understood. Now there is a major industry devoted to modifying certain molecules of biology in the quest for new properties.

Many of the molecules of modern experience, however, have been synthesized for the first time by synthetic chemists or by enterprising chemical engineers making sensible substitutions. Butyl rubber, dynamite, sulfuric acid, organic dyes, many drugs, polystyrene and chlorofluorocarbons are examples of useful molecules not readily available in our environment. (The starting materials for many useful molecules, of course, are derived from nature, perhaps from the bark of a willow tree or the berry of a suitable plant.) Many other molecules, such as those found in fuels such kerosene and gasoline are fairly easily made by fractionating heavy crude oils from natural deposits on which our energy economy strongly depends.

The task of the synthetic chemist is to provide the proper conditions so that the desired molecule will form. It may take many steps in sequence before the final product can be formed with a large yield. But the formation is still self-assembly, from our point of view.

Nanophysics and Nanotechnology: An Introduction to Modern Concepts in Nanoscience. Edward L. Wolf
Copyright © 2004 WILEY-VCH Verlag GmbH & Co. KGaA, Weinheim
ISBN: 3-527-40407-4

As elements of nanotechnology, molecules have some desirable characteristics. They are small enough, in many cases, to be unique and identical and free from errors, as are atoms. Quality control can be absolute in cases of small molecules such as water, diatomics, and even for relatively large molecules of high symmetry, such as the C_{60} Buckyball. (For all but the most esoteric practical purposes one can overlook the isotopic mass variability that does exist in most atoms and molecules.) This occurs in even the smallest atoms; atomic hydrogen comes in masses of 1,2,3, and helium comes in masses of 3 and 4, for example. The astounding success of biology as a nanotechnology has come with no need for isotope separation.

On the other hand, thermal excitations can lead to changes in shape of molecules on a rapid time scale. The thermal vibrations in cases of diatomic molecules and in the vibration of atom positions in extended solid are well understood and predictable in terms of thermal expansion and heat capacity, for example.

In contrast, devices built in the semiconductor technology are all different, in their complete atomic specifications, and only the essential parameters for the operation of the device are the realistic targets for quality control considerations. Devices of the most rudimentary sorts built up by assembly using scanning microscope tips are very challenging to build, and even more challenging to build in identical versions. The scanning tunneling microscope, as used in such cases as shown in Figure 3.8, does offer the opportunity to inspect in detail for errors and to correct them.

6.1
Carbon Atom $^{12}_{6}C$ $1s^2\,2p^4$ (0.07 nm)

The carbon atom is the basic unit of organic chemistry, biological molecules, notably DNA, the code for all terrestrial life. The six protons and six neutrons of the $^{12}_{6}C$ nucleus, self-assembled either in the big bang or in the interior of a star ($\sim 10^6$K). It is believed that supernova explosions have ejected nuclei and electrons into space, and these have gradually condensed in the atomic, molecular, liquid and solid matter of planets such as those in our solar system. The six electrons of the structure $1s^2 2s^2 2p^2$ can be expected to uniquely self-assemble about the $^{12}_{6}C$ nucleus on a very short time scale at almost any terrestrial temperature $<10^3$K. The radius of the 2p wavefunction is of the order of 0.1 nm. The outer 3s and 2p wavefunctions, amenable via hybridization to tetrahedral and other symmetries, easily bond into molecular and periodic solid structures, notably (metallic) graphite and (insulating) diamond, and, or course, organic chemistry and biology.

There is great flexibility in the way carbon bonds with itself and with other elements. The flexibility must be inherent in the wavefunctions as enumerated in Table 4.1, for these are well known to be the correct basis for all covalent bonding of carbon. In more detail, s and p_x combine for the sp bond of angle 180°, as in acetylene, C_2H_2; s, p_x, and p_y combine to form the trigonal sp^2 bond of angle 120°, as in ethylene C_2H_4; and s, p_x, p_y and p_z combine for the tetrahedral sp^3 bond of angle 109° in methane, CH_4, and diamond.

6.2
Methane CH$_4$, Ethane C$_2$H$_6$, and Octane C$_8$H$_{18}$

Methane is a covalent tetrahedral molecule involving the $2s2p^3$ electrons of carbon. The wavefunctions are listed in Table 4.1. In forming the bonds pointing to the corners of the tetrahedron, linear combinations of the 2s and 2p wavefunctions are selected. This is called hybridization of the underlying wavefunctions. The result is the filled $1s^2$ shell and the four orbitals pointing in tetrahedral directions, which are composed of linear combinations of 2s and 2p wavefunctions.

Ethane C$_2$H$_6$ involves the same carbon electrons. The carbons bond to each other and also to three hydrogens, again in the tetrahedral directions.

Octane C$_8$H$_{18}$ has one linear structure and also a rolled-up structure. The bonding in the linear form can be thought of as ethane, with six additional –CH$_2$-groups inserted between the original two carbons. This structure is not at all rigid. Molecular simulations of octane at a temperature of 400 K are shown in Figure 6.1 after Drexler [1]. The time intervals between the succeeding conformations are about 10 ps. The energy estimates come from computer modeling, as carried out by Drexler [1]. The great variability in conformation of this molecule under thermal conditions qualifies the earlier statements to the effect that molecules, once formed, are unique.

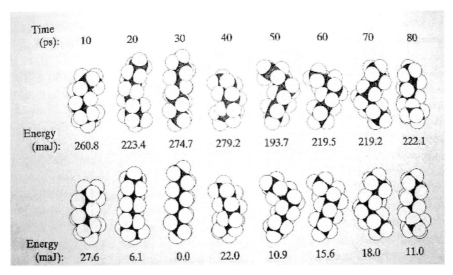

Time (ps):	10	20	30	40	50	60	70	80
Energy (maJ):	260.8	223.4	274.7	279.2	193.7	219.5	219.2	222.1
Energy (maJ):	27.6	6.1	0.0	22.0	10.9	15.6	18.0	11.0

Figure 6.1 Modeled thermal conformation changes of octane C$_8$H$_{18}$ at 400 K.
1 maJ = 10^{-21}J = 6.25 meV. From [1]

6.3
Ethylene C_2H_4, Benzene C_6H_6, and Acetylene C_2H_2

The carbon molecules considered so far have involved only single bonds, based simply on the covalent bonding found in molecular hydrogen that was discussed in Chapter 5, acting on wavefunctions as listed in Table 4.1. Carbon and other atoms also sometimes form double and triple covalent bonds.

The linear acetylene molecule C_2H_2 is formed by carbon atoms which each share three of their four valence electrons with each other, a structure called a triple bond. Thus, six valence electrons are concentrated in the region between the two carbons.

Ethylene C_2H_4 involves a double bond between the carbons, so each carbon devotes two electrons to its neighboring carbon and two electrons to hydrogens.

Benzene C_6H_6 involves a hexagonal ring of six carbon atoms. Each carbon is bonded to two carbon neighbors on the ring and to a hydrogen atom. The bonding between carbons is single on one side and double on the other, so that each carbon makes use of four valence electrons. The positions in the hexagonal ring at which the double- and single-carbon bonds appear, can be exchanged without changing the structure. This suggests that the carbon electrons involved in carbon–carbon bonding in benzene are delocalized and can move around the ring. There is evidence that this is the case from the large diamagnetic susceptibility of benzene. The six-fold carbon rings of benzene appear in graphite and its relatives, the carbon nanotubes and C_{60}. In the latter case there also five-membered pentagonal carbon rings, as well as the hexagonal rings.

Two benzene molecules sharing one side forms naphthalene, and three sharing sides in a linear array forms anthracene. The carbon atoms that are shared between two rings will have one double bond and now single bonds to two carbons. So the hydrogens on the interior of the 2D networks are released. The end point of this planar array of benzene molecules is the graphene sheet, and stacks of graphene sheets form graphite. This is one form of elemental carbon, involving only hexagonal rings with double and single carbon–carbon bonding. The bonding between sheets in graphite is weak and does not involve any covalent shared-electron bonds. The electrons in each sheet are delocalized making graphite a metal.

6.4
C_{60} Buckyball ~0.5 nm

These molecules are empty spherical shells, containing exactly 60 carbon atoms in five- and six-membered rings. It is essentially a graphene sheet closed onto itself, with pentagons added to allow curvature. The pentagonal and hexagonal benzene rings are located in a similar fashion to pentagonal and hexagonal panels in a soccer ball.

These molecules are very stable and fully tie up the four valence electrons of each carbon atom. A simple model of this molecule is shown in Figure 6.2 [2].

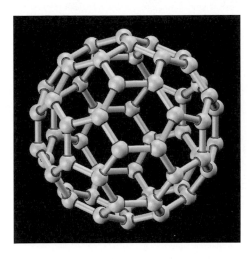

Figure 6.2 Carbon 60 molecule, in perspective model view. Each carbon atom participates in one double bond and in two single bonds. All valence electrons in each carbon are bonded. Each hexagonal benzene ring shares sides with three hexagonal and three pentagonal rings [2]

These molecules are present in tiny amounts in nature. They spontaneously form in oxygen-free atmospheres containing carbon atoms. There are reports of several other "magic numbered" molecules, notably C_{70}. It has been found possible to condense C_{60} molecules onto plane surfaces and to make superconducting compounds by adding stoichiometric amounts of alkali metals.

6.5
C_∞ Nanotube ~0.5 nm

Carbon nanotubes are cylindrical shells made, in concept, by rolling graphene sheets, choosing an axis vector, and a radius, and closing them. The diameters are in the range of 0.5 nm, similar to the C_{60} molecule, but the length can be micrometers. Once the radius and pitch are established, the tube will grow by adding carbon to its length, since the carbon bonds around the cylinder are fully saturated and do not offer an opportunity for additional carbon atoms to attach. The choice of rolling axis relative to the hexagonal network of the graphene sheet (the pitch), and the radius of the closed cylinder, allows for different types of single-walled nanotubes, which vary from insulating to conducting. There are also varieties of multi-walled nanotubes. These materials are very light and very strong. They self-assemble under various controlled conditions: by carbon arc discharge, laser ablation, chemical vapor deposition based on hydrocarbon gases, and laser assisted catalytic growth. The smallest diameter nanotubes, 0.5 nm diameter, have been grown inside zeolite cavities.

Figure 6.3 after Baughman [3] shows, in Panels A, B, and C, single wall nanotubes of three different types. Panel D shows a scanning probe microscope picture of a chiral single-wall nanotube, and Panel C shows a TEM image of a nine-wall nanotube.

The electrical conductivity of properly oriented (armchair) nanotubes is extremely high, with the observation that carriers can pass through the micrometer length of

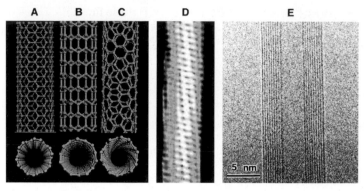

Figure 6.3 Images of carbon nanotubes [3]. (A), (B), and (C), respectively are armchair (metallic), zigzag (small bandgap) and chiral (semiconducting) nanotubes. The twist of the chiral nanotube is clearly evident in the lower panel (C), a perspective view along the tube axis. Panel (D) shows an STM image of a 1.3 nm diameter chiral nanotube. Panel (E) shows a TEM image of a nine-walled nanotube, a concentric cylindrically nested assembly, in which the binding between the adjoining nested tubes is very weak

available tubes with zero scattering, and zero heat dissipation. It has been reported [4] that a 1.4 nm diameter single-wall nanotube transmits 6 µA, corresponding to $> 10^8 \, \text{A/cm}^2$ (see Figure 6.3). Nanotubes frequently appear to be perfectly periodic over their lengths, on the scale of µm. The only source of electrical scattering is thermal vibration of the atoms. Vibrations have small amplitude in a rigidly bonded structure. In consequence, the electrical mobility is exceptionally high, which makes the units desirable for the active regions of high-speed electron devices.

Carbon nanotubes have also extremely high thermal conductivity, measured to be 3000 W/mK. This value is comparable to the best thermal conductors, diamond and basal plane of graphite, 2000 W/mK. The high thermal conductivity is significant for nanotechnology, because removing heat is one of the central challenges in making electronic devices smaller. Their tensile strength and Young's moduli are high, comparable to the strongest known materials (silicon carbide nanorods) and between 20 and 60 times stronger than steel wire [3].

Figure 6.4 [4] shows a 1.4 nm diameter single-wall nanotube assembled into a *field-effect transistor* (FET). The nanotube itself is regarded as self-assembled, but its incorporation into the electrical device using the methods of semiconductor technology is not yet self-assembly.

A challenge to the use of nanotubes in such detailed electrical applications is, indeed, how to get the right kind of nanotube (radius, pitch, wall thickness) into exactly the right location to do the task. Sorting of nanotubes can be approached on the basis of their electrical conductivity, highest for the metallic armchair tubes [see Fig. 6.3 (A)]. Another general approach is to grow the tube from a catalyst placed in an optimum position.

The nanotube based FET structure shown in Figure 6.4 [4] also functions as a light source. Electrons and holes are provided in the central portion of the nanotube by the separate source and drain contacts. Some electrons and holes recombine radiatively, radiating bandgap light. The bandgap of the nanotube (evidently a semiconductor) of about $E_g = 0.75$ eV, gives light emission of wavelength $\lambda = hc/E_g$ about 1650 nm. This is in the infrared, but not so different from the wavelengths involved in fiber optic communications. The length of the active region of the carbon nanotube, modeled as being between 400 and 800 nm, is large enough that this structure does not behave as a quantum dot. That is, the wavelength of the emitted light is determined by the nanotube's bandgap and semiconductivity (related to its pitch angle), and not by the length of the active portion of the nanotube.

The authors [4] report that the devices were fabricated by dispersing nanotubes in solvent, spreading the solvent on the lightly thermally oxidized surface (150 nm of gate oxide) of the P^+ Si gate electrode. After the solvent had evaporated, the nanotubes used in device fabrication were selected in a microscope. The source and drain and other supporting structures then had to be deposited using photolithography and methods of silicon technology.

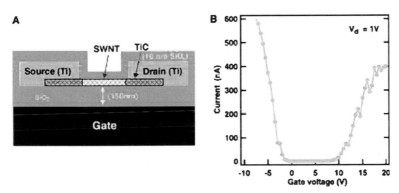

Figure 6.4 Incorporation of 1.4 nm diameter single-wall nanotube (SWNT) as active element of field-effect transistor [4] Panel (A) shows 1.4 nm diameter SWNT placed 150 nm above Gate (P^+ Si), spaced by 150 nm thermally grown SiO_2 dielectric barrier layer. Source and drain contacts to the nanotube are thermally annealed Ti, which establish covalent bonding to the C nanotube by the formation (during annealing) of titanium carbide. Electrically, the Ti-TiC-C contacts are highly transmissive Schottky barriers. The whole device is covered with 10 nm of SiO_2. Panel (B) Current–gate voltage characteristics at fixed source–drain voltage. Current is small except in positive and negative bias ranges when the nanotube becomes metallic in hole or in electron conduction. Saturation current of 600 nA corresponds to current density of $> 10^8$ A/cm^2

6.6
InAs Quantum Dot ~5 nm

Quantum dots have been mentioned in Chapters 1 and 4, in connection with fluorescent markers used in biological research. These are nanometer-sized single crystals grown in solution, taking care to give a uniform size. This type of self-assembly, to produce large numbers of nearly equal-sized products, is an accomplishment of modern chemistry and materials science. The basic idea is to let the crystals (or amorphous spheres, in the case of polystyrene beads) grow until the solution is depleted of dissolved atoms feeding the growth. When the supply stops the crystal stops growing, at a well defined size. In this scenario, there is also a need to control the number of nucleation sites.

Quantum dots in the application mentioned in Chapters 1 and 4 are useful because their controlled size, L, in turn controls the wavelength of light that is emitted when the crystal is illuminated by shorter wavelength light, sufficient to generate electron–hole pairs across the bandgap of the semiconductor. These quantum dots (QDs) are grown in solution, then coated in elegant ways, first, to protect the semiconductor crystal from an aqueous environment, and then, further coated to attach the fluorescent marker to the desired type of tissue.

Semiconductor (InAs) quantum dots (QDs) have found a rather different application in certain types of lasers. Lasers (Light Amplification by Stimulated Emission of Radiation) are widely used as optimum light sources in telecommunications, as well as in consumer products such as CD players. This QD application in telecommunications, again, is based on the *tunability* of the emission spectrum with the size, L.

However, these quantum dots are incorporated into a laser hetero-junction, in which electrons are injected from one side of the junction and holes are injected from the other side. The quantum dot is thus *electrically* "pumped" into an unstable electronic state from which it will radiate a photon, exactly as in the previous application. This is an example of an "injection laser", which is industrially important. But the question is "How can one grow quantum dots inside a P–N heterojunction"?

The answer is in the realm of *epitaxial film growth* on a single crystal substrate in a molecular beam epitaxy (MBE) machine, which is based on an ultra-high-vacuum (UHV) chamber. One example [5] is provided by InAs quantum dots, which can be made to self-assemble within a strained InGaAs quantum-well structure, grown epitaxially on (100) GaAs.

Epitaxial growth means that the material that is deposited (under ultra-high vacuum conditions, requiring suitable deposition rates and suitable substrate temperature) continues to grow on the same crystalline lattice as its substrate. Epitaxy may be preserved even if the chemical composition is changed, at the expense of some strain in the grown layer. For example, adding some Indium (In) to the Gallium (Ga) beam in growing GaAs on a GaAs substrate can lead to a perfect (but slightly strained) crystalline layer of IndiumGalliumArsenide (InGaAs), if the In concentration is not too large.

The effect of this Indium (In) addition to the growing layer, electrically, is to locally reduce the bandgap, E_g. Alternating layers of GaAs and (heteroepitaxial) InGaAs create a *square wave modulation of the bandgap, and this is called a quantum well structure.* The electrons will segregate into the layers having smaller bandgap (lowered conduction band edge), and these electrons are now in 2D bands (Chapter 4) because the thickness of the InGaAs layers is small. (This scheme raises the mobility, since the carriers collect in regions away from the donor ion charges, which would scatter them, and reduce their mobility.)

The quantum dot aspect appears if the In concentration is raised, increasing the local strain. Finally, the epitaxy will no longer be tolerated, and locally there will be *inclusions of InAs* (QDs). These are the "self-assembled InAs quantum dots", whose size (and emission spectra) are controlled by the details of the deposition conditions in this industrial process.

This is an advanced use of MBE technology to produce quantum dots (local deviations from epitaxy), which, in this case, make lasers vastly more efficient! Why? Because QDs have a sharp narrow range of light emission, much like atoms [5].

6.7
AgBr Nanocrystal 0.1–2 µm

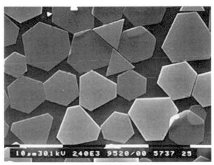

Electron micrograph of tabular
grain emulsion

Figure 6.5 Crystals contained in photographic emulsion are typically Silver Bromide (Image courtesy © Eastman Kodak Company. KODAK is a trademark.)

Crystals of AgBr and AgI are historically the basis for conventional silver halide photography. Figure 6.5 shows an array of such crystals in a tabular grain photographic emulsion. The range of sizes of such crystals is of the order of 0.1 µm – 2 µm. The process by which light absorption sensitizes such a crystal to allow it subsequently to be developed to a silver image (negative) is known in its outline but not in detail. It is likely to involve ruptures in the bonding of the AgBr crystal which lead to clusters of two or more Ag ions, which constitute a latent image, susceptible to the development process, in which relatively massive clusters of metallic silver are catalyzed by the latent image site.

Silver halide photography has historically accounted for a substantial fraction of the commercial sales of silver metal. The proper conditions for the AgBr crystals to self-assemble, and, certainly, the variety of subsequent steps in the category of "sensitization" to extend the spectral range of sensitivity, are known in the industry but not extensively in the open literature.

6.8
Fe$_3$O$_4$ Magnetite and Fe$_3$S$_4$ Greigite Nanoparticles in Magnetotactic Bacteria

Magnetotactic bacteria such as *Magnetospirillum magnetotacticum* are known to contain linear arrays of antiferromagnetic nanocrystals of Fe$_3$O$_4$ (magnetite) or Fe$_3$S$_4$ (greigite). The arrays are oriented parallel to the long direction of the bacteria. Both minerals are antiferromagnetic, which means that there are two magnetic sublattices of opposed but unequal magnetization, leading to a net magnetization in the direction of the stronger moment.

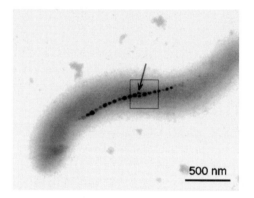

500 nm

Figure 6.6 Image of magneto-tactic bacteria, showing linear array of magnetic crystals. The nanocrystals are greigite, an iron sulfide, dimensions on the order of 40 nm. These crystals are 40 nm and are yet each large enough to be antiferro-magnets. Arrow in box locates a bi-crystal defect in the magnetic chain, see Figure 6.7 [6]

A typical bright field TEM observation [6] of a single cell of *M. magnetotacticum* (Figure 6.6) reveals a "magnetite chain 1200 nm long containing 22 crystals that have average length and separation of about 45 nm and about 9.5 nm, respectively".

The bacteria were cultured cells of Magnetospirillum magnetotacticum strain MS-1 and were deposited onto carbon-coated grids for use in the transmission electron microscope. The image was obtained at 200 keV with a field-emission-gun TEM. It was deduced that the linear array of the tiny magnets had a larger effect on the direction of the overall magnetization than did any magnetic anisotropy within individual grains. It is known in these systems that the crystals are combinations of (111) octahedron and (100) cube forms, and that the (111) magnetic easy axes of the crystals are primarily parallel to the chain axis [6].

Magnetic studies indicate that each crystal forms a single magnetic domain, and that the magnetization M in the magnetite is consistent with the bulk value, 0.603 T. The magnetic moment of one typical bacterium at room temperature was measured to be 5×10^{-16} Am^2. Measurements of the coercive field have given values in the range 300 Oersted to 450 Oersted. There are indications that smaller nanocrystals in the chains are sometimes superparamagnetic, rather than antiferromagnetic, and that their magnetization is stabilized by the magnetic field from larger neighboring antiferromagnetic nanocrystals in the chain. This may apply to the bi-crystal defect arrowed in the figure. These magnetic minerals evidently grow in the cell from iron-containing chemicals dissolved in the fluid inside the bacterium.

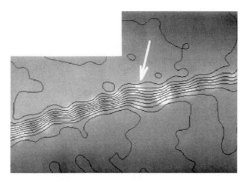

Figure 6.7 Magnetic field mapping [6] of central boxed region of magnetic bacterium image, shown in Figure 6.6 (see box and arrow). More precisely [6], this image is an off-axis electron hologram obtained in field free conditions. The apparent magnetic field lines thread along the array of antiferromagnetic nanocrystals, including the bi-crystal defect shown by the arrow, proving that the array acts as an extended bar magnet (believed to orient the bacterium in the earth's field)

These observations, of antiferromagnetism in 40 nm samples, add to the observations on the band gap, size-quantized excitation states, and tunable light emission of semiconductor quantum dots; as evidence that even very small samples may exhibit sophisticated phenomena for which the conceptual machinery of condensed matter physics is fruitful and necessary.

It is further believed that when these bacteria die, their magnetic moments remain intact and retain the orientation set by the magnetic field of the earth at the time of death. Over eons of time, extensive layers of sediment produced by these magnetic bacteria have accumulated. In taking core samples from the sea floor, changes in the direction of their magnetization as a function of depth have been recorded. It is believed that these data provide a record over long periods of time revealing the history of the orientation of the earth's magnetic field.

It is also known that magnetic crystals below 100 nm in size, occur in organisms in many biological phyla. Since these nanocrystalline magnets are apparently widespread in such primitive forms of life, one suspects that they may be present, at least in vestigial forms, in higher forms of life. If one does not know where to look in a large organism for such a tiny features, they may be hard to locate!

6.9
Self-assembled Monolayers on Au and Other Smooth Surfaces

Gold is a nonreactive metal, but sulfur-containing groups called thiols do have a tendency to weakly bond to a clean gold surface. Organic chain molecules, such as octanes, terminated in thiol groups, have been found to self-assemble on gold surfaces. They exhibit ordered arrays defined by their interactions with the surface, and with adjacent organic chain molecules.

The type of ordering that occurs in a given situation, when the thermal energy is low enough, is that which provides the lowest energy. For a two-dimensional array of spheres on a structureless surface, this is likely to be hexagonal close packing, so that each member of the array has the most benefit from the attraction it feels from its neighbors.

If there is a different kind of order in the substrate on which the array is deposited, for example, cubic ordering, then there will be a competition as to which interaction has the most to offer. If the spheres in the example have a strong attraction to the underlying cubic structure, then the adsorbed spheres may take on a cubic order to match. In cases of such competition, a remarkable degree of complexity can occur. Some of this has been documented in STM studies. It is even possible, in the case of two competing types of order, that domains of each type may coexist.

A good example [7], of an ordered monolayer, emphasized the locking of the monolayer structure to the structure of the underlying gold surface. This is not always the case. In the next example, there is no relation between the monolayer and the surface on which it rests.

Another example of an ordered self-assembled monolayer, shown in Figure 6.8 [8], is a hexagonal close-packed array of C_{60} molecules on an alkylthiol organic surface. The alkylthiol layer is similar to the layer in the previous example.

These authors emphasize that the hcp close packed layer, in which the C_{60} nearest-neighbor distance is 1.0 nm, is *unrelated* in its orientation to the rectangular lattice of the alkylthiols, which has dimensions 0.9994 nm by 0.8655 nm.

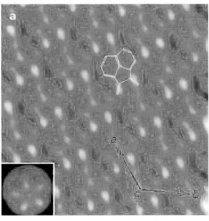

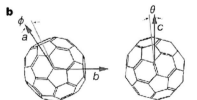

Figure 6.8 Self-assembled monolayer of identically oriented C60 molecules observed by cryogenic scanning tunneling microscope [8] Panel (a) reveals C_{60} array, individual molecules are identified by three resolved hexagonal rings (see white overlay in upper central field). Inset shows theory modeled image based on bond orientation in white overlay and axis identification acb in lower right. Panel (b) shows bond model of C_{60}: (left) oriented as it appears in image and (right) as it would appear if *c* axis is perpendicular to the monolayer

The measurements are made at 5 K in a cryogenic scanning tunneling microscope. Panel (a) shows a 3.5 nm × 35 nm scan of the field of C_{60} molecules. In this image, the structure of the Buckyball molecules can clearly be seen. The characteristic features are three hexagonal benzene rings which are clearly resolved by the STM image at 5 K. Further, and striking, is the fact that these three hexagonal features are seen to be locked in identical positions in all of the C_{60} molecules.

This gives an appreciation that molecules are absolutely identical, and in this case, identically oriented. The inset to the panel (a) shows a theoretical model of the STM image expected if the single and double bonds are oriented as shown (in white) in the center of the field. The agreement is seen to be excellent.

The bond models of C_{60} seen in panel (b) are oriented to match the images in panel (a).

The authors mention that in measurements at 77 K and at room temperature the images of the individual C_{60} are featureless. From this they infer that the molecules are in some state of rotation at 77 K and higher, but have locked into position at 5 K.

References

[1] From *Nanosystems: Molecular Machinery, Manufacturing & Computation* by K. Eric Drexler. Copyright © 1992 by Wiley Publishing, Inc. All rights reserved. Reproduced here by permission of the publisher.

[2] Courtesy Joseph W. Lauher, Chemistry Department, SUNY Stony Brook.

[3] Panels A, B, C and E reprinted with permission from R. H. Baughman, A. A. Zakhidov and W. A. de Heer, Science **297**, 787 (2002). Copyright 2002 AAAS. Panel D reprinted with permission from Nature: J.W.G. Wildoer *et al.*, Nature **391**, 59 (1998). Copyright 1998, Macmillan Publishers Ltd. and with permission from C. Dekker, Delft University of Technology.

[4] Reprinted with permission from J. A. Misewich, R. Martel, Ph. Avouris, J. C. Tsang, S. Heinze, and J. Tersoff, Science **300**, 783–786 (2003). Copyright 2003 AAAS.

[5] G. T. Liu, A. Stintz, H. Li, K. J. Malloy, and L. F. Lester, Electron. Lett. **35**, 1163 (1999).

[6] Reprinted with permission from R. E. Dunin-Borkowski, M. R. McCartney, R. B. Frankel, D. A. Bazylinski, M. Posfal, and P. R. Buseck, Science **282**, 1868–1870 (1998). Copyright 1998 AAAS.

[7] G. E. Poirier, Chem. Rev. **97**, 1117 (1997).

[8] Reprinted with permission from Nature: J. G. Hou, Y. Jinlong, W. Haiqian, L. Qunxiang, Z. Changgan, Y. Lanfeng, W. Bing, D. M. Chen, and Z. Qingshi, Nature **409**, 304 (2001). Copyright 2001, Macmillan Publishers Ltd.

7
Physics-based Experimental Approaches to Nanofabrication and Nanotechnology

The previous chapters have outlined the rules of nanophysics that apply to nanometer-sized objects. These rules describe chemical matter and biological matter. The previous chapters have also given some examples of nanometer-scale machines and devices achieved by nature and by modern science and technology.

It appears, to date, that the most advanced machines and devices on the nanometer scale are those produced in nature. The basic reason for this assessment is that nature seems uniquely to have developed an information base and locally-available instructions for assembly and replication of complex nanometer scale machines and devices.

On the other hand, the accomplishments of modern science and technology in two particular areas: the silicon and computer technology, and the scanning probe microscope approaches, exceed those available in nature.

Computation at gigahertz rates is not available in nature. Nor does nature allow individual atoms to be placed on selected sites to represent bits of information. That these accomplishments have come from essentially physics-related science and technology is notable, and encouraging to further efforts based on nano-physics.

On the other hand, it is not unlikely that the most immediately productive approaches to biotechnology may, indeed, be based on modifying natural (biological, genetic) processes to achieve practical goals, in the same spirit that fluorine substitutions for hydrogen have produced fluids, such as freon, more suitable for refrigeration. Genetically engineered corn is available, for example, which effectively generates its own insecticide. This is a contemporary accomplishment of biotechnology.

Of the many promising nanophysical approaches to nanotechnology, at present, a leading pair are: 1) Extensions of the hugely successful silicon/computer technology, and 2): Extensions of the methods of scanning probe microscopy. The scanning probe microscopy approaches, indeed, are closest, in a literal sense, to the imagined "molecular assembler" devices. There have been many examples, in the figures of this book, of the contemporary *silicon technology*, and of *scanning probe technology*,

These two areas have stimulated development of a large set of fabrication and measurement tools, which, in total or individually, may be useful as starting points for the invention of new forms of nanotechnology.

The continued advance of the silicon technology, it is argued, may lead to computers with capacity greatly exceeding the capacity of the human brain. A huge achievement (which has largely compensated its creators). *Is it also a threat?* When will these hugely intelligent computers and computer networks insist on a life of their own?

On the scanning probe technology side, advances may conceivably lead to realization of the "self-replicating molecular assembler", an imagined universal tool able to build "anything", quickly, on an atom-by-atom basis. What about this? If it is possible, how can one be sure that the "assemblers" might not run amok and use up all the chemicals in the world as we know it? Annihilate human life?

7.1
Silicon Technology: the INTEL-IBM Approach to Nanotechnology

Silicon microelectronics technology is a large topic. Only the barest outline of this material can be covered here. Several figures have already illustrated ingenious adaptations of silicon technology methods to the fabrication of hybrid structures; see, for example, Figure 6.4.

7.1.1
Patterning, Masks, and Photolithography

An essential element in silicon technology is preparing a sequence of patterns, miniaturizing the patterns, and transferring the miniaturized patterns sequentially onto the silicon chip to define the appropriate layers of metal (patterned into wires), insulation, and P and N regions needed for fabrication of logic and memory cells.

Patterning Deposition Masks
The patterning process starts with an image, for example, a detailed wiring diagram. This pattern has to be transformed from black and white (historically on drafting paper), to open and closed areas on a metal mask. The metal mask, which may be in the form of metal layers deposited on glass, would be open where the wire is to be placed. The size scale of the pattern may be 0.1 m on the drafting paper, but has to be reduced, say, to 1 μm on the contact mask. The full technology deals with single crystal wafers of, say, 6 inches diameter, containing many identical chips. There is a step-and-repeat aspect in the real process that we will pass over.

To utilize the mask, an initial layer of photoresist polymer, such as polymethylmethacrylate (PMMA), is uniformly applied to the chip surface. Then the mask is brought close to the coated surface, and light is exposed onto the photoresist through the mask. In this case the photoresist polymer would be of the type that is weakened by light exposure, so that the resist can be removed chemically, leaving the regions of the chip to be metallized ready for deposit of a metal layer, while the rest of the surface is protected by the photoresist. The whole surface is then coated with the metal for the wiring. Finally, in a chemical step, the remaining photoresist is removed, leaving the chip surface with a deposited wiring network.

Masking Layers to Limit Etching

Masking layers are used for other purposes than to define a deposit. Masking is also used to delineate areas of the surface that are to be removed by chemical etching or by dry etching. Dry etching is done in a plasma which may include reactive molecules such as CHF_3. (In many cases, these masking layers are also initially patterned by the basic deposition patterning outlined above.)

Masks to protect a surface are sometimes simply grown silicon dioxide SiO_2. Heating the silicon in oxygen or steam grows silicon dioxide layers of a wide range of thickness, up to many μm.

Alternatively, chemical vapor deposition (CVD) may be used to deposit a masking layer, such as Si_3N_4. In a CVD reactor the sample surface is exposed to chemical vapors, or carrier gases, which contain the chemical ingredients of the desired deposit. With control of the temperature of the surface and the flow rates and compositions of the carrier gases, CVD is a way to rapidly grow layers of a variety of compositions. For example, carrier gases silane and hydrogen at 630 °C can be used to deposit polycrystalline silicon (polysilicon), which may also be doped N-type with addition of a phosphorus-bearing chemical vapor to the reactor. Phosphosilicate glass can be deposited by CVD methods.

Metal layers are also commonly used as masks against etching, for example, gold and chromium. Extremely thin layers of Cr or Ti are also used in advance of a thicker layer of a conducting metal, to improve adhesion of the electrode to the silicon.

7.1.2
Etching Silicon

Wet Etches

Etching away silicon is accomplished by wet etches, which include acids HF (which removes oxide quickly and Si slowly), nitric and acetic acids. These are isotropic etchants.

An isotropic etch, when exposed to a surface protected by a planar oxide or metal masking layer, will have a certain tendency to undercut the masking layer, producing in a limiting case a spherical cavity. The shape of the etch pit can be controlled, of course, by limiting the etching time.

Still in the category of wet etches, the alkali KOH is an example of an anisotropic etchant, which removes material from certain crystallographic planes faster than from others. KOH can be used on Si to produce triangular etch pits exposing (111) planes on a (100) surface.

Dry Etches

Dry etches, protected by masking layers such as oxides or metals, are accomplished in a vacuum chamber or gas reactor. In this case energetic ions, produced in a plasma discharge, hit the surface and remove material. (Removal of material under bombardment by ions is called sputtering, and is also adapted as an important method for thin-film deposition.) The ions may be non-reactive, for example, argon ions, or reactive, such as CHF_3 ions. These etch processes are somewhat directional, because the momentum carried by the ions in the plasma is predominantly perpendicular to the surface.

Directional dry etching can be carried out using an ion-beam gun, which is an accessory to a vacuum chamber, and generates a directed beam of ions, which can be reactive or non-reactive. In this way, deep trenches with vertical walls can be produced using the conventional metal- or oxide-etch masks.

7.1.3
Defining Highly Conducting Electrode Regions

Transistors, bipolar and FET, require precisely defined localized regions of N- and P-type conductivity. An NPN transistor might be formed by diffusing two closely spaced N regions into a P-type surface. A field effect transistor (FET) requires highly conductive source and drain contacts to be closely located and connected by the channel, which is a high-mobility conductive region. The conductivity of the channel of an FET is modulated by an electric field provided by voltage bias on the gate electrode, which is insulated from the channel by a thin layer of oxide (the gate oxide).

The traditional means of producing N or N^+ and P or P^+ regions is by thermal diffusion from a dopant-containing layer positioned appropriately. (Doping can also be introduced by ion-implantation, see below.) The chemical dopant is contained in a glass, so that when the heating is carried out, the dopant does not simply evaporate. Typical dopants for N-type silicon are phosphorus and arsenic; for P-type silicon, boron is typical. Silicon is a robust (although very easily oxidized) material and temperatures above 600 °C are needed to produce significant diffusion. At these temperatures photoresist layers cannot be used, but oxide or metal masking layers are still appropriate.

Ion implantation is another method for introducing impurities and is described in the next section.

7.1.4
Methods of Deposition of Metal and Insulating Films

Evaporation
Metal deposition is traditionally accomplished by evaporation. Evaporation of a soft metal like Al can be accomplished by heating the metal in an electrically heated boat of Ta until it melts and until its vapor pressure becomes high enough to eject energetic atoms of Al toward the surface to be coated. This process requires a high vacuum system, typically evacuated with an oil diffusion pump, an ion pump, or a cryopump.

For more refractory metals, like Cr or Ti, evaporation is done using an electron-beam gun (e-gun). In an e-gun an electron beam is focused on the surface of the metal charge, locally heating it to a very high temperature, leading to evaporation as in the case of Al. Evaporation, including e-gun evaporation usually cannot be used to deposit compounds, because the different elements in the compound typically have different vapor pressures.

Sputtering

Sputtering is another basic technique for making deposits of almost any metal. In sputtering a discharge of non-reactive ions, such as argon, is created, which fall on the target and break loose atoms (and atom clusters) which are collected on the surface to be coated. Sputtering can also be used to make deposits of compounds. Different sputtering rates for the different elements of a compound may require a target composition somewhat different from that of the desired deposit, but this method for depositing compounds is much more suitable than evaporation.

Chemical Vapor Deposition (CVD)

Chemical vapor deposition is also a general method for depositing metals as well as compounds. In a CVD reactor, carrier gases, containing the elements of the desired compound, flow over the surface to be coated. This surface is heated to a suitable temperature to allow decomposition of the carrier gas and to allow mobility of the deposited atoms or molecules on the surface. Mobility of deposited atoms helps to produce a highly ordered crystalline deposit, desired for improved electrical properties. In some cases laser light may be used as an assist (LA-) in the decomposition of the carrier gas. Processes called LACVD are used to produce nanowires, for example.

Laser Ablation

Laser ablation is a general method for depositing even a complicated compound from a target to a nearby surface in vacuum. Focused laser pulses fall upon the target, completely vaporizing a small region of material. The ablated material is collected on the nearby surface, and has essentially the same composition as the target. This is a slow and expensive process by its nature, but finds application in high-value products where a complicated compound, such as a high-temperature superconductor (HTS) with many different elements, must be deposited with accurate stoichiometry. Incorporation of HTS films as wavelength filters in cellular telephone transmitter output, can multiply the number of mobile telephone subscribers who can be served by a given cellular phone transmitting station, for example.

Molecular Beam Epitaxy

Finally, molecular beam epitaxy (MBE) is an elegant process for depositing atomically perfect (epitaxial) layers of compound semiconductors, such as GaAs and InGaAs. This is an expensive and slow process, but is still industrially competitive in cases where the product is valuable.

In MBE, separate ovens with careful temperature regulation are provided for each element in the desired deposit. Temperature control is needed to control the vapor pressure and hence the deposition rate. The rates of each element have to precisely match the stoichiometry. For InGaAs, e.g., there will be three ovens. Deposition rate monitors and shutters will be available for each oven. With these elaborate precautions, in ultra high vacuum, epitaxial layers and heteroepitaxial layers can be deposited.

For example, multiple quantum well devices are produced using MBE. These methods were described in Chapter 6 in connection with "quantum-dots in a quantum well" lasers, valuable for the improvements and economies they may offer in optical-fiber communications or even in lasers for portable CD players, where battery life may be an issue.

Ion Implantation

High-energy ion beams can be used to implant chemical impurities such as donor and acceptors. An ion implanter is an electrostatic accelerator capable of giving 50 – 100 keV energy to ions, which allows them to penetrate deeply, on a micrometer scale, into a single crystal of silicon. The violent ion implantation produces undesirable damage to the host crystal, which would drastically lower the mobility for current flow, but annealing (heating) can remove the implantation damage. The temperature required for annealing (removing) implantation damage is lower than that required to diffuse impurities into the silicon from a surface layer of, e.g., phosphosilicate glass. For a large-scale contemporary silicon device facility, ion implantation is a standard tool.

Ion implantation can be used to create a *buried* layer of impurity, including the buried oxide layer (BOX) used in prototype FET transistors on silicon. In this case oxygen ions are highly accelerated and a heavy dose is buried at micrometer depths into the silicon crystal. After annealing, the result can be a nearly perfect silicon layer (Silicon on Insulator, or SOI) of perhaps 4 – 8 nm depth above a thick SiO_2 layer, above the single crystal Si wafer.

7.2
Lateral Resolution (Linewidths) Limited by Wavelength of Light, now 180 nm

7.2.1
Optical and x-ray Lithography

The widths of the finest (narrowest) wires that can be put on a chip, in optical lithography, are limited by optical diffraction. The light that exposes the photoresist comes through an aperture defined by the metal walls of the mask. Light, as a wave, has a tendency to bend around a corner, and this means that the sharpness of the image of the mask on the photoresist, is limited by λ, the light wavelength.

At present, this limits wires to being at least 180 nm. There is no easy way around this limitation, because higher energy (shorter wavelength) light sources are inconvenient and expensive. In addition, the higher energy light, which can extend to the x-ray range, then also has a tendency to penetrate the conventional metal mask. Metal wires, if they were made much narrower than 180 nm, still retain their essential characteristics as conductors, although the resistance per unit length increases. This incremental effect will increase the amount of heat per unit length, on a constant current basis.

180 nm is a lot bigger than the 1.4 nm diameter nanotube of Figure 6.3, which carried a current of 0.6 μA. On the other hand, there is no known prospect at present for creating the correct number of nanotubes of the right radius and pitch, and to place them appropriately on a 6 inch diameter wafer.

These issues, among others, definitely limit the advance of the traditional "INTEL-IBM" silicon technology. This performance is conventionally compared, in the industry, to the numerology of "Moore's Law" (see Figure 1.2). The current performance and projections for semiconductor technology are summarized on a timely basis at "The International Technology Roadmap for Semiconductors" [1], available on the Web at public.itrs.net/.

7.2.2
Electron-beam Lithography

Photolithography is the conventional (hugely successful) process for make computer chips *in bulk*. Finer lines can be produced by an alternative process, electron beam lithography, but this process is not compatible with the mass production available with photolithography, where the masks are simultaneously illuminated with light, exposing many chips at once.

In the e-beam process, an electron beam is focused to a point, as in an electron microscope, to individually expose the photoresist. The diameter of an electron beam can be as small as the resolution of an electron microscope, approaching 0.1 nm. However, the electron beam has to be scanned, as in a TV picture tube, to illuminate every area of photoresist. *This is a slow process.* Consequently, e-beam lithography is used in research, but not in production, in applications to produce features of dimensions smaller than 180 nm.

7.3
Sacrificial Layers, Suspended Bridges, Single-electron Transistors

We have already seen, in connection with our discussion of the Casimir force, in Chapter 5, a structure in which a flat plate was freely suspended above a silicon surface by two torsion fibers. This structure (see Figures 5.5 and 5.6) is an example of one in which a "sacrificial layer" (in this case silicon dioxide) was initially grown.

A deposited *polysilicon layer*, suitably patterned, was grown above the sacrificial layer, also providing supports to the underlying crystal. Then the SiO_2 layer was etched away, leaving the suspended "paddle" free to rotate in response to electric and Casimir forces. The method is seen in Figure 7.1 [2].

Referring to Figure 7.1, panel (1) shows e-beam exposure of a PMMA (polymethylmethacrylate) resist, which has been uniformly applied above the three layers comprized by the base silicon, the sacrificial oxide layer, and a polysilicon deposited layer. Panel (2) shows a patterned metal etch mask on the polysilicon layer. Panel (3) shows liftoff of the unexposed resist layer, followed in panel (4) by dry etch to delineate the suspended bridge. In panel (5) a wet etch, such as HF, undercuts the oxide beneath the suspended layer, leaving it freely suspended. Finally, in panel (6), metal electrodes are applied.

This scheme is used in the class of MEMS (NEMS) devices Micro (Nano) Electro-MechanicalSystems, which include the accelerometers used in automobile collision

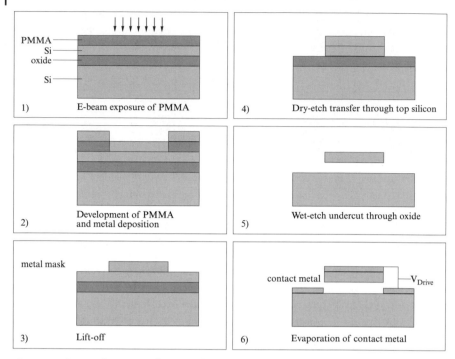

Figure 7.1 Steps in formation of a suspended silicon plate above a silicon chip [after 2]

sensors. It is also used in the recent "Millipede" prototype data storage device, where more than a thousand individual atomic force microscope sensors are produced, using the silicon photolithographic process illustrated in Figure 7.1, on a single chip [3].

Another example of this elegant version of the silicon process is illustrated in Figure 7.2, in which a single-electron-transistor (SET) is used to sense the 0.1 GHz vibrations of a suspended nanometer-scale beam [4,5].

The structure in Figure 7.2 panel (b) is patterned and etched from a GaAs-GaAlAs single crystal heterostructure. The vibrating beam is a doubly clamped single crystal of GaAs with dimensions 3 μm × 250 nm wide × 200 nm thick, with a resonant frequency of 116 MHz. The beam (the gate electrode of the SET) was patterned [5] using electron beam lithography and a combination of reactive ion etching (to make the vertical cuts) and dilute HF etching (to remove the sacrificial GaAlAs layer originally under the beam). The beam, which serves as the gate electrode, is capacitively coupled to the Al island, the closest approach being 250 nm, with a gate-island capacitance of 0.13 fF.

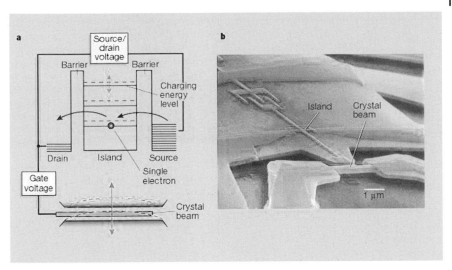

Figure 7.2 Single-electron-transistor used to sense vibration of freely suspended crystal beam [4,5]. In panel (b) a "crystal beam" is suspended and free to vibrate at 116 MHz. It is metallized and acts as gate electrode in the FET transistor, denoted "island". The "island" is connected to source and drain electrodes (right and left) by Al-AlOx-Al tunnel barrier contacts. At fixed gate bias and fixed source-drain voltage, motion of the gate electrode induces charge on the island, and sensitively controls the source-drain current. Panel (a): Schematic diagram of electrical operation of the device. A motion of the bar by 2 fm (about the radius of an atomic nucleus) can be detected with 1 Hz bandwidth at 30 mK

7.4
What is the Future of Silicon Computer Technology?

The ongoing improvement of silicon computer chips continues, with innovative new device designs appearing frequently. In addition to the International Technology Roadmap for Silicon [1], a further analysis of the trends is given by Meindl *et al.* [6]. One of the recent trends has been to build the basic field effect transistor, FET, or MOSFET (Metal Oxide Silicon Field Effect Transistor), on a silicon chip with a "buried oxide" layer (BOX). The effect of the BOX is intended to reduce the thickness of the active channel, which carries the current when the device is in the "ON" (conductive) condition. One advantage of the reduced channel thickness is to reduce capacitance between the source and drain. Reducing this capacitance allows the device to be switched more rapidly, with less energy wasted per switching cycle.

Figure 7.3 shows a *prototype* field-effect transistor described by IBM [7] as the "world's smallest working transistor". It is an example of "silicon on insulator" technology, with a silicon thickness of 4–8 nm. The silicon channel length, L_{Gate}, is 6 nm. Some features of this device are produced lithographically using light of 248 nm wavelength.

It seems clear that the gate, because of its extremely short length, can only be produced by e-beam lithography, an unconventional and slow method. Further, there is

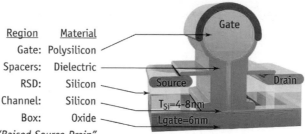

Region	Material
Gate:	Polysilicon
Spacers:	Dielectric
RSD:	Silicon
Channel:	Silicon
Box:	Oxide

RSD stands for "Raised Source Drain"

Box stands for "Buried Oxide"

Figure 7.3 Schematic indication of MOSFET *prototype* of extremely small dimensions, based on the conventional silicon technology [7]. Looking at this Figure, starting from the bottom: The bottom layer is "Buried Oxide" (Box), and at the top of this thick oxide layer, in a small thickness T_{Si} of 4 – 8 nm, is the active silicon channel. The other notably small dimension in this device is the length L_{Gate} of the gate electrode, stated as 6 nm. This extremely small dimension implies the use of e-beam lithography, certainly a complicating factor

no mention of the gate oxide, which is one of the most difficult features of the basic MOSFET design in connection with maintaining the trend as set by Moore's Law. The problem that occurs with scaling is that the gate oxide becomes so thin that it is no longer a barrier, as required in the framework of an FET device. Barriers so thin as to allow nanophysical tunneling, inevitably the consequence of reducing the size scale, are incompatible with conventional FET operation.

An innovative feature in this design is the thinning of the active silicon channel by means of the buried oxide layer, or "Box". One way to form the Box is to heavily implant oxygen ions deeply into a silicon crystal, followed by annealing. The details (of the implantation and anneal) control the channel thickness T_{Si} and determine whether the active channel is depleted or partially depleted.

It appears that the buried oxide method is preferred over the more obvious approach of depositing the thin active silicon layer onto a sapphire chip. (Sapphire is a good substrate because of its extremely high thermal conductivity.) The quality, including mobility and reproducibility, of the resulting thinned silicon layer is presumably higher in the Box approach.

This energy-saving approach (and the publicized use of copper, rather than aluminum wiring, for its modest improvement in conductivity) is a hint that the heat generated in the chip is recognized as a limiting factor in the scaling of the silicon technology.

7.5
Heat Dissipation and the RSFQ Technology

The generation of heat in silicon chips is a problem at present, and is a problem that will get more difficult the more devices there are on a chip [1]. Large installations, as well as the uncomfortably warm contemporary laptop computers, already exhibit this problem.

There is a complete, *but problematic*, potential solution to this heating problem, suitable for large installations where expensive air conditioning is already required. The potential solution is the *Rapid Single Flux Quantum (RSFQ) superconducting technology*. Although the full details of this technology, purely based on nanophysical phenomena of superconductivity, are beyond the scope of this book, it is a wonderful example of the notion that the new laws of nanophysics may offer concepts for different, and possibly superior, devices.

An example, the performance of an RSFQ counting circuit [8] is shown in Figure 7.4.

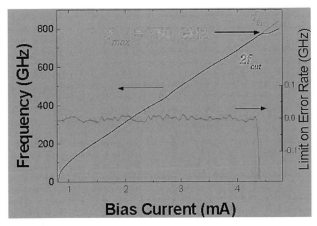

Figure 7.4 Frequency divider [8] operates at 750 GHz, using RSFQ superconducting junctions. This device produces half-frequency output. Twice the output frequency, $2f_{out}$ is shown superimposed on f_{in}, both quantities referenced to left-hand scale. Bias current (abscissa) in mA does not signify heat, because the resistance is zero [8]

The problem, for the RSFQ technology, is that refrigeration is needed to maintain RSFQ devices at the temperature of liquid helium. Liquid helium is a coolant widely used in Magnetic Resonance Imaging (MRI) medical installations. In addition to saving energy, the RSFQ technology allows *extremely fast clock speeds*, as seen in Figure 7.4.

The heat dissipation that occurs during the RSFQ switching events is described [8,9] as on the order of 10^{-18} J/bit, leading to an overall power dissipation estimated as 10^{-5} that of an equivalent silicon device. These devices are fast, as can be seen by the graph. These RSFQ technology devices have been considered for analog-to-digital converters, e.g., in missiles where *fast* data acquisition is urgently desired and refrigeration is already in place to cool infrared detectors.

The RSFQ technology is relatively simple, tolerating larger linewidths, but suffers from a primitive (underfunded) infrastructure for fabrication and testing. The active devices are niobium-aluminum, niobium (tri-layer) [10,11] Josephson tunneling junctions, which must operate well below the superconducting transition temperature of niobium, 9.2 K. Compact low power refrigerators that approach this temperature are becoming available.

The energy consumption and the cooling power (note that these quantities are additive, in terms of operating costs) needed for a large computer in this technology are less, by large factors, than for an equivalent silicon (CMOS) machine. The savings in power and floor-space are significant. It can be argued that the size of the overall machine would be much smaller because the design would not have to have the fins, open channels, fans, and air conditioners that are normal in large silicon computer/server-farm installations, simply to keep the devices from dangerously overheating, and to carry away the heat.

A benchmark in large-scale computing is the Petaflop computer, which carries out 10^{15} floating point operations per second. A typical modern Pentium-chip laptop computer operating at 1.3 GHz has a power supply rated at 54 Watts. A rough conversion of 1.3 GHz clock speed to 1 GFlop, would imply that a million laptop computers, if suitably coupled, might constitute a parallel architecture Petaflop computer. (This practical approach to making supercomputers has been implemented, as mentioned in the press, including use of Pentium computers and also SONY PlaystationII devices).

On this basis, a Petaflop computer would consume 54 MW of power, or 1.7×10^{15} J/year. At a conventional rate of 10 cents per KWh, the annual power operating cost would be $47 M, not to mention the cost of the air-conditioning.

Binyuk and Likharev [9] offer the following projected comparison of the specifications for a Petaflop computer (10^{15} operations/s) in RSFQ technology, operating at 5 K, vs. silicon CMOS technology operating somewhat above room temperature:

In CMOS, one might have [9] $10^4 - 10^5$ advanced CMOS silicon chips, total power about 10 MW. Neglecting the air-conditioning needs, the operating cost would be around $8.7 M/y. The linear size of the computer might be 30 m, leading to a relativity time delay $L/c = 300$ ns. Such a large signal-propagation time is a drawback for a fast computer. This is a huge installation, on the scale of those in existence at major stock exchanges.

In RSFQ, one might have [9] 500 RSFQ logic chips and 2000 fast superconductor memory chips, at an estimated power at 5 K of about 1000 W. The linear size of the computer CPU might be 1 m, corresponding to a speed-of-light time delay of 20 ns. To achieve the reliable cooling of this 1 m "core" to about 5 K, a closed-cycle (recycling) liquid helium refrigerator would be required, at a total power of about 300 KW. Again neglecting air-conditioning, the annual cost is estimated as $263 000 per year.

This is a noticeably lower figure than $8.7 M/y estimated for the CMOS facility, to which one should add much larger costs in floor space and air-conditioning.

The ever-increasing energy cost of the pervasive silicon information technology was hinted at in the media discussion of the US power blackout of August 14, 2003. There were statements to the effect that the US needs an (electric) power grid that "matches the needs of information technology". "Server farms", typically housed in urban buildings with windows blocked in, that have appeared in information technology, are users of vast amounts of electrical power.

This costly power shows up as heat in the environment, cost on the balance sheet, and even represents a potential security threat, in its dependence on large amounts of (demonstrably insecure) electrical power.

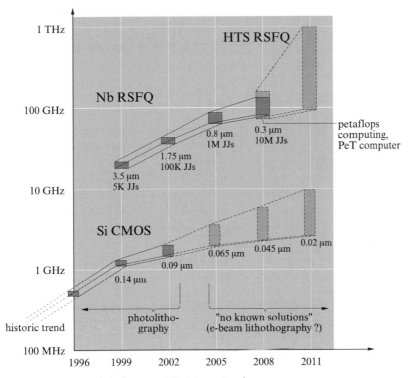

Figure 7.5 Estimated clock rates in RSFQ (upper) and conventional Si CMOS (lower) projected to the year 2011 [after 9]

Figure 7.5, adapted from Binyuk *et al.* [9], shows comparative projections of clock rate in RSFQ and CMOS computer technology. In the upper portion, "Nb RSFQ" refers to technology based on Nb-Al-Nb trilayer [10,11] Josephson junctions (JJ's), which operate below 9.2 K. In the upper right, "petaflops computation" and "PeT Computer" ("Personal Teraflop Computer", see [9]), with 10 million JJ's on a single chip, are predicated on an assumed 0.3 μm Nb technology. Note that 0.3 μm exceeds today's silicon technology linewidths, and is likely achievable with funding. In the far upper right, "HTS RSFQ", is a rough guess of RSFQ performance if the high-temperature superconductors could be brought into play.

The HTS superconductors are notoriously difficult to incorporate into electronic device technology, except in the simplest single-layer applications, such as high-Q radio-frequency filters for cellular telephone transmitting stations. For HTS superconductors the required refrigeration to 77 K is less costly.

In the lower portions of Figure 7.5 relating to the silicon technology [1], note that the linewidths by 2003, have already come down to the limit of photolithography. The very uncertain projections labeled "no known solutions" reflect that no mass production technology has appeared to replace the photolithography. These projections are based on the Silicon Technology Roadmap [1].

Other large issues relevant to Figure 7.5 are the greatly reduced power consumption in the RSFQ technology, coupled with its extremely inconvenient cooling requirements. These issues have been discussed in the text above.

It is clear that RSFQ technology will never find its way into wristwatches, laptop computers and most other basic computing venues for the silicon chip. The RSFQ technology may find commercial application in high-end installations requiring massive rapid computation.

The RSFQ technology offers a chance for much reduced power consumption and better performance in computing technology, at the major cost of learning how to make and use a variety of good refrigerators and making a large adjustment in mental orientation from silicon to superconductors like niobium.

To challenge and surpass a pervasive and successful technology, such as the silicon technology, is seldom achieved, and certainly not quickly. Jet airliners now outnumber propeller driven planes, and digital cameras may be displacing conventional silver halide photography. On the other hand, in spite of the complete success of nuclear power in the US Navy for submarines and aircraft carriers, private enterprise in the United States regulatory environment seems unable to make nuclear power work.

The RSFQ approach [9] offers an opportunity for innovators and entrepreneurs who are conversant with the advanced, extensive, and almost wholly untapped nanophysical science of superconductivity. To launch this radical computer technology, which is certainly technically sound; requires a largely new workforce and a start-up approach aimed at a niche market. There is a small core of dedicated workers competent in superconducting technology, including RSFQ, in laboratories at universities and a few leading defense contractors.

7.6
Scanning Probe (Machine) Methods: One Atom at a Time

The scanning tunneling microscope (STM) and the atomic force microscope (AFM) are similar single-point probes of a surface. Both have a sharp tip which can be moved in an x-y raster scan over a 2D surface and both have a servo loop which raises the height of the tip to maintain the current (or the surface-tip force) to be constant, thereby providing a topograph of the surface. Either instrument can be imagined as capable of carrying an atom or molecule to a point on the surface and leaving it to provide one step toward fabricating a surface structure. Either can be imagined as a starting point for the "machine assembly" of a nanostructure, atom by atom. The essential ideas of the two instruments are presented in Figure 7.6 [12].

The STM is capable of atomic resolution, as is evident from many images, including those in Figure 7.7. The basic idea is that the tunneling occurs when the wavefunction of an atom on the tip overlaps the wavefunction of an atom on the surface. In ideal cases these wavefunctions decay exponentially with characteristic distances on the order of the Bohr radius, around 0.1 nm. With such a short decay length, a single tip atom may carry almost all of the observed current. If the wavefunction on that atom is, e.g., a directed bonding orbital, the spatial resolution may be remarkably high, as is sometimes observed.

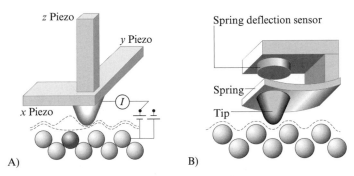

Figure 7.6 Schematic diagrams of (A) scanning tunneling microscope (STM) and (B) atomic force microscope (AFM). For STM the tunnel current *I* is maintained constant by tip height, controlled by z-piezo. For AFM the force (spring deflection) is maintained constant [after 12]

The maximum scanning and sampling rates are determined in large part by the resonant frequencies of the support structure, the tip and the *xyz* piezo-electric element on which it is mounted. The upper range of such frequencies in the best instruments is about 1 MHz.

The AFM (see panel (B) in Figure 7.6) is more complicated. Sensing of the force between tip and surface is by deflection of the cantilever on which the tip is mounted. The sensing in modern AFM instruments is by deflection of a light beam, focused on the upper surface of the cantilever; or by changes in the resistance of the specially designed cantilever with deflection (piezoresistance). The earliest instruments sensed the deflection of the cantilever by changes in the tunneling current at fixed bias between the tip holder and an upper spring-deflection-sensor electrode (see panel (B)).

The force between the tip and sample is attractive at large spacing (van der Waals regime) and repulsive at small spacing (exclusion principle overlap). Atomic imaging has been reported with the AFM, but is more difficult than in the STM.

7.7
Scanning Tunneling Microscope (STM) as Prototype Molecular Assembler

7.7.1
Moving Au Atoms, Making Surface Molecules

A recent state-of-the-art STM study, which combines atomic resolution imaging and nanofabrication using an STM, is indicated in Figures 7.7, 7.8 and 7.9. [13].

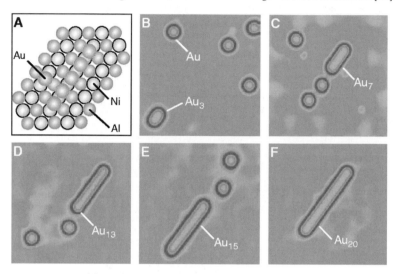

Figure 7.7 STM fabrication and atomic imaging of gold Au atoms and molecules Au_n on a surface of (110) NiAl [13]. The NiAl (110) surface has dimpled channels matching the 0.3 nm spacing of Au atoms in Au molecules. A 20-atom Au_{20} wire is formed and its density of states matched to a 1D band model (see Chapter 4)

Figure 7.7, panel (A) schematically shows the (110) surface of the NiAl crystal on which Au atoms were randomly deposited. (B) 9.5 nm by 9.5 nm image shows 4 Au atoms and an Au_3 cluster observed in STM at 12 K, using the constant current mode at 1 nA. The sample is at positive 2.1 V bias, which means that electrons from the tip tunnel to empty states in the Au 2.1 eV above the Fermi level. The STM tip is used to pull the atoms into line in panels (C,D,E,F), leading to the 20-atom nanowire.

Pulling the Au atoms is accomplished by decreasing the tunneling resistance to 150 KOhms, compared to 1 GOhm for observation. The higher electric field, at reduced resistance, induces a larger electric dipole in the Au atom, which is then attracted to the tip, where the field is strongest. Pulling is a balance of forces between the Au atom and the tip and surface. The Au atom is neither attached to, nor lifted above, the surface by the tip.

Figures 7.8 and 7.9 show tunneling spectroscopy measurements and their analysis along the 20-atom nanowire. These measurements, and their analysis, are useful in preparing for a further step in nanofabrication [14], in which these authors react the Au nanowire with a CO molecule. The bonding between the Au wire and the CO depends upon the electronic structure of the Au wire, which is analyzed here.

In the theory of electron tunneling, the conductance, dI/dV, is a measure of the density of states per unit energy [15]. Positive bias voltage here signifies the energy relative to the Fermi energy of the empty state in the wire that is being filled by the tunneling electron.

The dI/dV–V curves taken near the middle of the wire show a large peak at 0.78 eV. This peak is easiest to think of as the singularity in the density of states at the bottom of a 1D band, as described in Chapter 4. It also represents the $n = 1$ state of the electron trapped in the 1D potential well of length L. The bottom of the empty band is determined as $E_o = 0.68$ eV above the Fermi energy. The details of the dI/dV–V curve reflect the fact that the system is a 1D electron trap of length L. Thus, the wavefunctions of the states that the electrons can tunnel into are of the form $\sin(n\pi x/L)$, where x is measured from one end of the wire. It is determined that the principal values of n involved (see Figure 7.9, panel (A)) are 1, 4, 5, 6, and 8. The width of the levels is about 0.35 eV, corresponding to a lifetime of about 10^{-15} s.

It is seen that the states overlap in energy, so that in the density of states measurement at 0.78 V (Figure 7.9, panel (C), upper curve), the electrons fall partially into

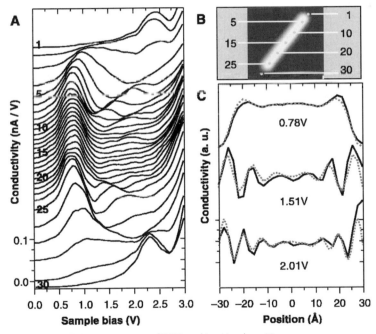

Figure 7.8 STM measurement of DOS vs. bias V and position along the Au_{20} wire. V denotes energy of empty state where tunneling occurs, and dI/dV (nA/V) measures DOS. Panel (A): curves of dI/dV at locations seen in panel (B). Panel (C) shows DOS vs. x along the wire, taken at three different energies [13]

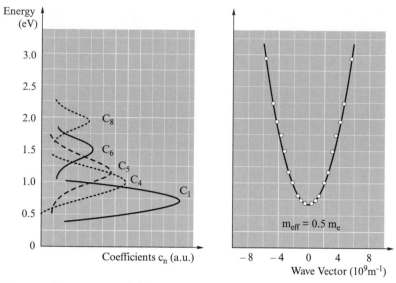

Figure 7.9 Determination of effective mass of electrons in the one-dimensional nanowire [after 13]. The effective for electrons in the nanowire is determined to be about 0.5, from the *E–k* curve shown in Figure 7.9, panel (B).

the $n = 1$ state and partially into the $n = 4$ state. The dotted curve shown in comparison to the solid measured curve is based on this idea, and fits quite nicely. The more rapid oscillations in the 1.51 V and 2.01 V curves of Figure 7.7, panel (C), are explained by the higher n values appropriate at these these energies, and the oscillatory nature of the probability density $\sin^2(n\pi x/L)$.

This experiment and its analysis can be interpreted as again showing that the concepts of solid state nanophysics, such as the energy band, and the effective mass, are useful even for tiny systems. In a 20-atom system, the set of curves in Figure 7.8 panel (A) is interesting in this regard. The curves taken with the tip near the center of the sample show the characteristic of a 1D energy band as described in Chapter 4, with a huge peak at the bottom of the band. The curves taken at the ends of the 20-atom wire, however, deviate from the essentially bulk behavior seen in the middle. This illustrates an important point in nanophysics, that the smaller the system, the larger are the edge- or surface- effects. The reason is, simply, that the surface (or end or edge) of a small system is a larger fraction of the total than it would be of a large system.

One aspect of the situation not shown in these figures is observation of the bonding and antibonding states of the Au_2 molecule [13]. The single gold atom (not shown) shows a large density-of-states peak at 1.95 V, which is understood as a prominent empty state in the atomic density of states of an Au atom on the NiAl surface. However, in the Au_2 molecule, as in the H_2 molecule studied in Chapter 4, this state splits into two states. For the Au_2 molecule, the two states [13] are symmetrically located above and below the 1.95 V state, at 1.5 V and 2.25 V. This splitting indicates a strong interaction between the adjacent atoms at their spacing of about 0.289nm, spacing similar to the spacing of Au atoms in bulk gold.

From the point of view of an STM as a prototype "Molecular Assembler", this work is also instructive. The 20-atom gold wire is assembled by the tip, *after the atoms have been randomly deposited* on the carefully prepared surface at a very low temperature. The low temperature is necessary to keep the atoms from randomly diffusing about, because their trapping into the dimples of the NiAl surface is weak. The tip does not *carry* the atoms into the nanostructure to be assembled, but merely nudges existing atoms into position. The process illustrated in Figure 7.7 can be assumed to be extremely slow.

7.7.2
Assembling Organic Molecules with an STM

Chemical reactions have been carried out on single crystal surfaces of copper by Hla *et al.* [16], as illustrated in Figure 7.10 [17]. The upper and lower panels show molecules lined up along a monoatomic step on the surface, which offers stronger bonding than does a flat surface. The starting molecules are iodine-substituted benzene rings, namely iodobenzene. In the first inducement by the tip, a voltage pulse breaks off the iodine, leaving a phenyl ring. The tip is then used to pull the iodine out of the way, and then to pull two phenyl rings close together. The second inducement provides energy to allow the two adjacent phenyl rings to fuse and form $C_{12}H_{10}$, the desired product. Finally, in panel (f), pulling on one end of the biphenyl is demonstrated to move the whole molecule.

This experiment shows a new use of the STM tip, in adding energy to either dissociate a molecule into parts or to help two molecules to react, by providing energy to overcome a barrier to the reaction. In this experiment a specially prepared Cu surface was used, one cut at a slight angle to form monoatomic steps (shown as shaded horizontal strips in the upper and lower sections of Figure 7.10). This step-edge location offers stronger binding to an atom or molecule, because two sides of the atom are affected, rather than one, on a single flat surface.

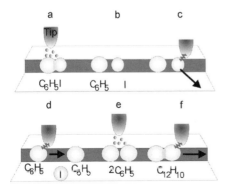

Figure 7.10 Schematic of single molecule STM organic chemistry on a copper surface [16,17]. The reaction is the copper-catalyzed transformation of iodobenzene C_6H_5I to make $C_{12}H_{10}$. This reaction is known as the Ullman reaction, here carried out at 20 K, with the usual thermal activation energy replaced by pulses of energy from the tip. The first pulse is applied in panel (a) to dissociate the I, leaving a phenyl group C_6H_5. The second pulse is applied in panel (e) to cause two adjacent phenyl molecules to bind. In panels c, d, and f the tip is used to move a molecule

7.8
Atomic Force Microscope (AFM) Arrays

The atomic force microscope is useful for making marks on surfaces, which can be described as nanolithography, as well as for measuring topography of surfaces that need not be conductive. An example of an AFM image is Figure 4.5 revealing several InP nanowires on a flat surface. The AFM tip can be used to push and drag objects like nanowires or nanotubes around once they have been deposited on a flat surface. The AFM has been adapted to work in liquid environments in biological research. As a prototype for the "molecular assembler" the AFM tip is quite similar to the STM tip.

In neither case does the single tip provide an advanced and controllable facility for picking up, manipulating, or specifically releasing an atom or molecule. To assemble a structure using LEGO toys, the hand, under a watchful eye, selects a part, grasps that piece, orients the piece, inserts the piece into the desired site, and then releases the piece. *These functions are essentially unavailable with STM and AFM tips.* The smallest radius that can be expected for an AFM tip, mass produced in silicon technology, is likely to be 20 nm [3].

Additional functions that an STM/AFM tip system can provide include a combination of current and force sensing, as would be available from a piezoresistive, electrically conductive cantilever. An added ability [3] is for tip heating and tip temperature sensing. In [3] the tip temperature was used as a means of reading digital information imprinted into the surface. However, there is no indication in the present literature suggesting that there will ever be a means of grasping and releasing an atom or molecule from an STM or AFM tip.

7.8.1
Cantilever Arrays by Photolithography

The Millipede project [3], carried out by IBM at its Zurich Laboratory, provides a 32×32 array (1024 tips) contained in an area of 0.3 mm $\times$ 0.3 mm, and capable of writing and reading 10 Gb of information with access times on the order of 25 ms. The information is recorded by thermal-mechanical indentation of 40 nm size pits in a re-writable thin PMMA polymer film.

Conceived of as a data storage device that might replace the magnetic disk, the Millipede achieves data storage density of 400–500 Gb/in^2. This density is far beyond present disk memory and also beyond the reliably estimated "super-paramagnetic" limit for magnetic disk storage, around 60–100 Gb/in^2. This device, which has access times similar to disk memory and operates in air at room temperature; has been built, tested, and quite fully described.

The technology of Millipede builds upon the previously described methods for creating suspended beams or bridges, illustrated in Figures 7.1 and 7.2. A 32×32 field of partially detached cantilevers is provided on the "cantilever chip". The cantilevers are constructed entirely of crystalline silicon, including scanning tip, tip heater, and temperature sensor. The cantilever chip and the closely matched PMMA-coated "recording chip" are fabricated separately, and are subsequently mounted together using a delicate x,y,z drive which moves *all cantilevers in unison* relative to the recording chip. The cantilevers are not individually moved in x,y, or z. The writing and sensing of information by each cantilever are accomplished, respectively, with the tip heated under indentation force, and, for reading, by the individual temperature sensor.

The dimension of a cantilever cell is $92\,\mu m \times 92\,\mu m$. Each cantilever has two legs, which join at the tip-tip-heater assembly. The legs are doped silicon: $50\,\mu m$ long, $10\,\mu m$ wide and $0.5\,\mu m$ thick. The resonant frequency of each cantilever is $200\,kHz$ and the force constant is $1\,N/m$. Each of the 1024 lithographically patterned Si tips is $1.7\,\mu m$ in height, triangular in shape with an apex sharpened to a $20\,nm$ radius using oxidation sharpening [3]. The tip heater time constant is a few μs, which is expected to give a rate of $100\,kHz$ for reading and writing. The tips are heated to 500–$700\,°C$ for writing (making an indentation in the PMMA polymer), and to around $350\,°C$ for thermally sensing the presence of a data point (indentation). Provisions are available for erasing data points.

7.8.2
Nanofabrication with an AFM

The Millipede device described above is intended as a memory device, but we can think of its writing function as a step of nanofabrication or molecular assembly.

Ignoring all questions about the facility of a single $20\,nm$ radius tip to do the traditional functions in mechanical assembly, let us focus on how fast the assembly could possibly be carried out.

The Millipede stores $10\,Gb = 10^{10}$ bits with an access time of $25\,ms$. We can optimistically translate that as a rate of 4×10^{11} steps/second, which is a great overestimate. (It is a great overestimate because each tip can access only 1, certainly not all, of its own $10^{10}/1032 = 10^7$ sites in $25\,ms$.) Suppose the goal is to fabricate a diamond structure of mass $12\,g$. The molar mass of carbon is 12, so the product will contain Avogadro's number, 6.02×10^{23}, of atoms. Even at the overstated rate of 4×10^{11} steps/second, the required time will be 1.5×10^{12} seconds. *This is a very long time, about 48 000 years.*

To fabricate the 12 gram carbon structure in $1\,s$ would require 1.5×10^{12} Millipede devices working at once, again overlooking other absurdities of this idea.

For a second estimate, assume a single assembler tip at the highest possible rate, about $1\,GHz$. *At that rate, the time for a single tip to process 12 grams of carbon would be 6×10^{14} seconds, 19 million years.*

These rates are so slow, using even wildly optimistic estimates on tip rates and capabilities, as to make clear that *bulk matter cannot be assembled in an atom-by-atom mode* except by huge numbers of "molecular assemblers" working in parallel. The assemblers themselves must be very small and probably have to grow in numbers by reproducing themselves, as do the cells in biology.

This is the point in the argument for molecular assembly where the further assertion is made that the assemblers self-replicate. Following an illustrative analysis provided by Smalley [18], assume that the "molecular assembler" can reproduce itself at the same rate of 10^9 atoms per second. Let's further assume that the "molecular assembler" comprises only 10^9 atoms. On this assumption it will ideally take one second to make one copy of itself; if each copy reproduces itself in turn, then in 60 seconds the number of "molecular assemblers" will be 2^{60}, or about 10^{18}! (This is pure fantasy, but it is similar to the fantasy upon which the nanotechnology myth of "gray goo" rests.) If all of these huge numbers of assemblers subsequently work at the same rate, 10^9/s, then carbon can be assembled at a rate of 10^{27} atoms/s! For comparison, this is $10^{27}/(6.02 \times 10^{23}) = 1661$ Moles/s $=20$ Kg/s for molar mass 12 g for carbon. This is a large and industrially relevant rate, but it is *impossible*, since it is based on the flawed assumption of the molecular assembler.

This is not at all to say that arrays of AFM tips are useless for data storage, quite the contrary. It is clear that the Millipede device exceeds the projected limiting performance of any possible magnetic disk memory. The dollar value of disk memory in today's commerce is vast. Thus, a new memory technology is anticipated, based on nanophysics and nanotechnogy.

7.9
Fundamental Questions: Rates, Accuracy and More

We have looked at the most advanced results from the leading scanning probe devices. These devices have produced new understanding of the behavior of matter at the atomic scale. *As far as possible use as "molecular assemblers" on an atom-by-atom basis for bulk matter, there is nothing that we have seen that remotely suggests that this is possible.*

The single tip, however refined it may possibly become in extra facility for grasping, orienting, and releasing particles, simply cannot act rapidly enough to produce a measurably large sample in a useful period of time.

Arrays of tips, projected [3] in the Zurich IBM Lab to rise to a million tips on a chip, do not come close to filling the rate discrepancy that is inherent in the large size of Avogadro's number.

References

[1] *The International Technology Roadmap for Silicon*, on the Web at www.public.itrs.net/.

[2] H. G. Craighead, Science **290**, 1532 (2000).

[3] P. Vettiger, M. Despont, U. Drechsler, U. Durig, W. Haberle, M. I. Lutwyche, H. E. Rothuizen, R. Stutz, R. Widmer, and G. K. Binnig, IBM J. Res. Develop. **44**, 323 (2000).

[4] Reprinted with permission from Nature: M. Blencowe, Nature **424**, 262 (2003). Copyright 2003 Macmillan Publishers Ltd.

[5] R. G. Knobel and A. N. Cleland, Nature **424**, 291 (2003).

[6] J. D. Meindl, Q. Chen, and J. A. Davis, Science **293**, 2044 (2001).

[7] Courtesy: IBM Research, T. J. Watson Research Center. Unauthorized use not permitted.

[8] Reprinted with permission from W. Chen, A. V. Rylakov, V. Patel, J. E. Lukens and K. K. Likharev, Appl. Phys. Lett. **73**, 2817 (1998). Copyright 1998, American Institute of Physics.

[9] P. Bunyk, K. Likharev, and D. Zinoviev, Intl. J. of High Speed Electronics and Systems **11**, 257 (2001).

[10] M. Gurvitch, M. A. Washington, and H. A. Huggins, Appl. Phys. Lett. **42**, 472 (1983).

[11] E. L. Wolf, *Principles of Electron Tunneling Spectroscopy*, (Oxford, New York, 1989) pp. 215–230.

[12] P. K. Hansma, V. B. Elings, O. Marti, and C. E. Bracker, Science **242**, 209 (1988).

[13] Reprinted with permission from N. Nilius, T. M. Wallis, and W. Ho, Science **297**, 1853 (2002), published online 22 August 2000 (10.1126/science.1075242). Copyright 2002 AAAS.

[14] N. Nilius, T. M. Wallis, and W. Ho, Phys. Rev. Lett. **90**, 186102 (2002).

[15] E. L. Wolf, op. cit., p. 317.

[16] P. F. Schewe and B. Stein, Physics News Update (AIP) Number 503, September 22, 2000.

[17] S. W. Hla, L. Bartels, G. Meyer, and K. H. Rieder, Phys. Rev. Lett. **85**, 2777 (2000).

[18] R. E. Smalley, Scientific American **285**, #3, 76 (2001).

8
Looking into the Future

It is clear that nanophysics will increasingly contribute to the development of technology as the decreasing size scales require quantum concepts. Those working in the various related areas of materials science, silicon technology, device design and fabrication will benefit by understanding the quantum phenomena which will inevitably show up on the smallest size scales.

An example from the foregoing is the quantum dot, where very useful behavior, a size-dependent shift in wavelengths absorbed and emitted, is adequately explained starting from the simplest quantum concept, the trapped particle in one dimension. Exploiting this uniquely nanophysical behavior has contributed to development in biological research and in laser development, as well as in electron device technology.

On a practical basis, it should be clear that a more detailed understanding of the basic phenomena of nanophysics, a topic not often emphasized in contemporary college science and engineering in the US, can indeed be relevant and profitable to a worker interested in nanotechnology. The conceptual thread of nanophysics underlies the properties of atoms, molecules, solids, electron devices, and biological science, including notably its instrumentation.

Nanophysics also bears on some conceptions, expectations, and misconceptions of nanotechnology. Nanoscience touches on some of the larger issues of human existence as well as on smart ways to make new machines and new industries. Nanotechnology has even been depicted as a threat to life. What about this?

8.1
Drexler's Mechanical (Molecular) Axle and Bearing

Tetrahedral bonding as occurs in diamond, silicon and germanium is the basis for Drexler's proposed family [1] of "diamondoid" nanostructures. These are large and complicated, covalently bonded, molecules. Start with a molecular axle and bearing (sleeve bearing) as shown in Figure 8.1. There is little doubt that this structure, and others like it, would be stable and robust, as calculated, *if it could be formed*.

The suggestion seems to be that such molecular structures could be designed *and eventually built* to form nano-replicas of common mechanical devices starting with

Nanophysics and Nanotechnology: An Introduction to Modern Concepts in Nanoscience. Edward L. Wolf
Copyright © 2004 WILEY-VCH Verlag GmbH & Co. KGaA, Weinheim
ISBN: 3-527-40407-4

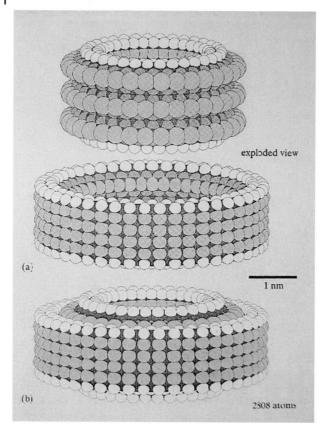

exploded view

(a)

1 nm

(b)

2808 atoms

Figure 8.1 Proposed sleeve bearing composed of 2808 diamondoid atoms [1]

elementary bearings; planetary bearing, pumps, and building up to molecular scale
robotic devices, machines, and factories.

8.1.1
Smalley's Refutation of Machine Assembly

As has been clearly pointed out by Smalley [2], such structures (e.g., see Figure 8.1)
cannot be fabricated using STM tips. Richard E. Smalley, who won the Nobel Prize in
chemistry for his work in discovering and fabricating C_{60} and carbon nanotubes, is
an authority on nanotechnology. There is no answer to Smalley's basic point, that
these large single molecules have interior locations that are completely inaccessible
to such a relatively large object as an STM tip. Smalley also points to the inadequate
functions of any such single tip, suggesting that several tips might have to work
together to achieve the needed functions, making the tip size even larger and more
of a problem. The total inadequacy of the rate of atom-by-atom assembly for making

bulk matter has been considered earlier, and is also discussed at length by Smalley [2].

Additional arguments support Smalley's position, including the tendency, noted in connection with Figure 3.9, for an inserted atom to choose its own preferred bonding configuration, rather than the one required by the minimum energy for the overall structure.

Another way of stating this point is to note that the structure in Figure 8.1 has many strained bonds, to achieve its curvature. If the atoms were put into place in sequence, the deposited atom would choose an unstrained bond, tending toward a straight beam rather than a beam bent into a circle.

These structures cannot be assembled atom by atom [2]. On the other hand, structures such as this presumably stable molecule, might self-assemble, in the traditional mode of synthetic chemistry, in the right environment. A good environment for growing a structure must allow modifications of the structure to be tested for their relative stability. This means that the thermal energy must be sufficiently high to allow alternative structures to form, at least on a transient basis. A rough measure of that thermal energy is the melting temperature of the type of solid structure that is involved. The melting temperatures of strong solids are in the range 2000 K – 3000 K. To grow a diamondoid structure such as Figure 8.1 would require a temperature well above 1000 K, an environment completely devoid of oxygen, probably rich in hydrogen, and containing stoichiometric amounts of the elements for the particular structure. The structure in Figure 8.1 is stated to contain 2808 atoms.

Synthetic diamonds are commercially available, and are used in abrasives and other applications. Their formation requires high pressure and high temperature. Physicists and materials scientists have had a difficult time in controlling the growth of these small crystals, in any shape. Synthetic chemists have an honorable and productive history and a good intuition as to what kinds of molecules can be formed and how. However, synthetic chemists have not notably solved the problem of making diamondoid molecules or solids.

A step in this direction has recently been reported by Scott *et al.* [3], in an article entitled "A Rational Chemical Synthesis of C_{60}". The process involves 12 steps from commercially available starting materials by rational chemical methods. The final step is "flash vacuum pyrolysis at 1100 °C" (1373 K). Here again, the pyrolysis shows the need for a really energetic environment to close such a strongly curved structure. It appears that C_{60} is the only product in the reported synthesis. The authors comment that the approach they have used "should make possible ... preparation of other fullerenes as well, including those not accessible by graphite vaporization". The decomposition temperature of graphite is stated as 3300 K.

To make graphite from heavy hydrocarbons requires extended exposure to an extremely high temperature in a reducing atmosphere. To make a C_{60} molecule or carbon nanotube, conventionally requires enough energy to produce elemental carbon, as in a carbon arc, or ablated carbon plasma, corresponding to a temperature certainly in excess of 1000 K (the decomposition temperature of diamond is listed as 1800 K, and of SiC is 2570 K). The outstanding stability and relative simplicity C_{60}

explains its presence, along with C_{70} and other relatively stable species, as the atomic carbon condenses from a discharge.

In the condensation of a high-temperature, oxygen-free, atomic vapor containing the correct amounts of diamondoid elements, the single-molecule wheels and gears of Drexler (similar but more complicated than the sleeve bearing of Figure 8.1); or, perhaps, variants of such designs containing fewer strained bonds, will certainly have a chance of forming. That chance, however, may well be negligibly small.

Catalysts, if such can be found, would certainly improve the prospects of certain species. It is known that once the radius and twist of a nanotube is established, for example, on the surface of a tiny catalyst, perhaps a nano-sphere of carbide-forming metal, then the tube tends to grow rapidly along its length. Similar linear growth seems to occur for silicon, and other nanowires (See Figure 6.3 for an InP nanowire). A nanotube contains only one element, however, while the typical envisioned wheels and gears contain a range of covalently bonding elements.

In summary, the diamondoid structures cannot be produced on an atom-by-atom basis [2]. The extremely strong bonding and chemical variety in these structures makes their synthesis very difficult, except in cases of the fullerenes, which have only one element and very symmetric structures. Diamond-like carbon molecules of arbitrary shapes would also seem to be inaccessible to synthesis, as certainly are the diamondoid molecules of reference [1].

8.1.2
Van der Waals Forces for Frictionless Bearings?

Frictionless motion seems almost attainable, between flat or regularly curved surfaces which are nonreactive, attracted by van der Waals forces and repelled by atom–atom overlap forces. The best examples are graphite, in which the individual planes rather easily slide past each other, and in nested carbon nanotubes. In the case of nested nanotubes, translation and rotation have been carefully observed. These surfaces are non reactive (no dangling bonds) and the spacing is so small that no foreign atoms can exist in the gap. To the degree that the surfaces are commensurate, one would expect a locking tendency, but with weak attractive forces this may be negligible. In theoretical work [4] non-commensurate surfaces have been suggested to minimize any tendency to locking. These theoretical ideas can be tested against measurements on nested nanotubes.

8.2
The Concept of the Molecular Assembler is Flawed

The molecular assembler and the "self-replicating molecular assembler" are ideas that have been much discussed. From the point of view of nanophysics, such devices are not possible.

The molecular assembler is imagined to perform the functions of a hand or a robot arm on single atoms. The proposed device selects a particular kind of atom, grasps the atom, orients it suitably, inserts it into an atomic site in a molecule or solid being assembled atom-by-atom. It is necessary for the atom in question to be detached from the assembler arm and deposited in the proper site.

Molecular assemblers have been presented as able to assemble large machines out of diamondoid elements on reasonable time scales. Such projections must assume vast numbers of assemblers, considering the rate analysis discussed in Chapter 7. But, no matter, for no such tip-like device working on single atoms is possible. Any such tip is too large to allow access to atomic sites in a complicated structure. A small enough tip does not have facility for orienting an atom and for adjusting its property to grasp and then to release the atom. The rate at which any such device could operate is too slow to be of use in producing gram quantities of matter on an atom-by-atom basis. The device is said to work on any atom and to build any structure.

The argument advanced as an existence proof for molecular assemblers is nature itself, which we know has complicated and diverse molecular scale assembly capabilities. *Yet, there is nothing in nature that remotely resembles the imagined molecular assembler, in form, appearance, or general philosophy.* In nature, very specific enzymes apply to very specific molecules, to catalyze their formation or to cut them at particular locations. The idea is more like a lock and a key. Rather than working on atoms, the assembly processes in nature, work from a large inventory of molecules and polymers, which are formed (self-assembled) in traditional modes of chemistry and polymer chemistry. Rather than being general, and applying to a wide variety of structures (as the molecular assembler is imagined to do), the key, or decision-making, assembly processes in nature are extremely specific. There is an extensive hierarchy of such assembly processes, but they are independent and individually they are extremely specific, working on entities much larger than atoms.

Look at Figure 5.4, showing, in outline, the process of splitting a double helix DNA strand into two identical strands (the DNA replication fork). In this figure, all of the constituents: the bases A, C, G, and T; the deoxyribose cyclic sugar molecule, and the ionic triphosphate; are self-assembled molecules resulting from well-known biochemistry. *Even nature does not attempt to build anything atom by atom.* The decision-making "assembly process" in this figure is extremely specific, following from the lock-and-key fit of only two specific base-pairs across the helix from one strand to the other. Only the A will fit the T in bridging the strand, and only the C will fit the G.

To conclude, from the point of view of nanophysics, the molecular assembler is not possible. This of course means that self-replication of such imagined objects cannot occur. Nanotechnology (as differentiated from biotechnology) is therefore not to be feared from the point of view of possibly generating self-replicating organisms. These alarming ideas have come from a basically flawed concept and can be rejected as impossible.

While it is conceivable that entirely different chemical environments might allow the evolution of distinct forms of biology and life, no chemical environment will allow the imagined molecular assembler. This can be rejected simply on the properties of atoms. Atoms are the same throughout the universe.

What can occur, and has occurred to some extent already, is that great understanding and control of molecular biology allows modification of viruses, genes, and forms of life. Everyone agrees that these are topics in genetic engineering and biotechnology.

8.3
Could Molecular Machines Revolutionize Technology or even Self-replicate to Threaten Terrestrial Life?

In view of the discussion above, as long as "molecular machines" are understood to be the "assemblers" which apply to the diamondoid elements previously discussed, the answer to the question is "no".

Unfortunately, there is a history of misplaced concern that needs to be addressed. The field of nanotechnology has suffered from alarming warnings from its inception. The early popular book "Engines of Creation" [5] starting from an assumption of "replicating assemblers ... able to make almost anything (including more of themselves) from common materials", goes on to warn that "dangerous replicators could be ... too rapidly spreading to stop" and refers to the scenario of imagined replicators spreading like a cancer as the "gray goo problem" (p. 172). This attention-getting notion seems distinctly misleading from the present view of nanophysics, in which there is absolutely no chance of molecular assemblers ever existing apart from biology and plausible variations on biology.

This alarming possibility [5] was recently recalled [6] by Bill Joy in an essay entitled "Why the future doesn't need us". Joy mentions "genetics, nanotechnology and robotics" (GNR) as three dangerous areas of research that prudently should be banned.

Fukuyama, in his book "Our Posthuman Future" [7] (primarily a rational analysis of ethical issues of genetic engineering), casually links "nuclear weapons and nuclear energy ... perceived as dangerous from the start ..." to "nanotechnology – that is, molecular-scale self-replicating machines capable of reproducing out of control, and destroying their creators." Both categories, he says (p. 8) are "threats ... easiest to deal with because they are so obvious." The answer for nuclear weapons and nanotechnology, according to Fukuyama [7], is strict control on an international basis. Fukuyama, dismissing nuclear power and nanotechnology as being taken care of by strict control, goes on to consider genetics and biotechnology in careful detail.

From the point of view of nanophysics, a *link of nuclear weapons and nanotechnology is ludicrous*. There is no particular danger in small-sized particles of matter. Fear of nanotechnology is not justified. Such fears appear to be based on the imagined, fallacious, but widespread notion of the "self-replicating" molecular assembler.

The imagined danger of the "molecular assemblers" of nanotechnology is a red herring, for no such assemblers will ever be possible. The assembly processes of nature, of course, do allow the spread of dangerous organisms, in the form of viruses and bacteria. "Genetic engineering" and "biotechnology" of viruses and bacteria, and even of higher life forms, do, certainly present threats as suggested by Bill Joy [6]. Genetic modifications and biotechnology are real and active areas of research and production that have major present and potential impacts, including the possibility of a "posthuman" future.

Nanotechnology, as distinct from genetics, biotechnology and robotics, is a natural inter-disciplinary area of research and engineering based on nanophysics, chemistry, materials science, mechanical and electrical engineering and biology, and has no particular risk factors associated with it. Nanometer-sized particles of all sorts have been part of nature and the earliest forms of industry for all of recorded history. Nanometer-sized chemical products and medicines are routinely subject to screening for safety approval along with other chemicals and medical products.

8.4
What about Genetic Engineering and Robotics?

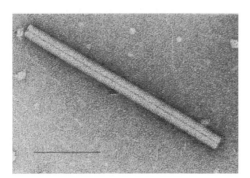

Figure 8.2   TEM image of the Tobacco Mosaic Virus (TMV) [8]. Note scale marker 100 nm. The TMV is 18 nm by 300 nm. This is the earliest identified virus, not a living object, but capable of parasitically entering living cells. Great facility now exists worldwide for modifying such viruses

Figure 8.2 shows a TEM image [8] of the Tobacco Mosaic Virus, the first virus to be identified. A virus is not a living object. Rather it is a chemical structure which is capable of entering an appropriate living cell, parasitically altering the operation of that cell, and ultimately replicating within the host organism. Viruses are the causes of many diseases, including polio and anthrax.

The New York Times recently reported [9] the experience of John Tull, a native of Santa Fe, New Mexico, who was infected by bubonic plague, apparently by a flea on his own property in suburban New Mexico, in the year 2002. Mr. Tull became gravely ill while visiting New York City, but his life was saved, at great effort and sacrifice on his part, in New York City hospitals. The effects of the (initially very few) *Yersinia pestis* bacteria on Mr. Tull's system were so great that the doctors had to induce Mr. Tull into a coma for some weeks, and, finally, to amputate his feet, simply to keep him alive.

Great facility now exists, all over the world, in the production, alteration, and application of viruses. Modified viruses can be used to produce modified cells including modified versions of bacteria such as *E. coli* and *Yersinia pestis*.

The New York Times of August 19, 2003 reports [10] a medical procedure on a Parkinson's disease patient. The patient's skull was opened to allow insertion of *an artificial virus*. The virus was a naturally available virus, which had been modified (engineered) to carry a specific DNA sequence, which in turn was designed to counter the effects of Parkinson's disease. This experimental procedure had been approved by the Federal Drug Administration for 12 patients with advanced Parkinson's disease.

Research workers at State University of New York at Stony Brook [11] reported synthesis of the polio virus from starting chemicals available in the modern biological laboratory. The only starting information used was knowledge of the DNA sequences in the polio virus. The implication is that this virus could be built to altered specifications, perhaps including those for which the Salk vaccine would be useless.

A large industry exists today based on biological molecules with modified properties. Enzymes are central to the biological breaking down of large molecules into smaller molecules. Enzymes have been modified and successfully introduced into common household detergents with the purpose of breaking down dirt and difficult stains. Corn seed is widely available which has been genetically modified to provide, in effect, its own pesticide. Considerable furor greeted the potential introduction of seed corn genetically engineered in such a way that the resulting seed would be infertile. Marketing of this product, engineered with the so-called "terminator gene", eventually was abandoned.

The perceived [6] threat from robotics and artificial intelligence is related to the hugely increasing computing power, and the continuously advancing field of artificial intelligence.

A corollary to Moore's Law (see Figure 1.2) is the notion that at some not far-distant time the computing power of a computer may exceed that of a human being. This comparison does not make sense in many ways, because the type of computation in the human and in the silicon chip is so different. There is a push to find ways to make these two types of intelligence more compatible, to provide a "broadband connection" to the brain. Scenarios along this line are considered at length by Kurzweil [12]. The idea that, somehow, a huge computer system might become willful (The Age of Spiritual Machines), is very intriguing. Whether this is possible is not obvious, but some scientists expect it may eventually be. It is certainly assumed in recent Hollywood films such as The Matrix and Terminator.

On the one hand, any self-empowered computer system will presumably be constrained to live in its air-conditioned box, limiting the damage it might inflict. The option would seem to exist to pull the plug. On the other hand, a huge parallel-processing system controlling large sums of money in a bank, if it were to undergo a phase transition and declare independence, might, in its own territory, be able to do a lot of financial mischief.

A recent and thorough analysis and projections in the robotics/artificial intelligence area have been given by Rodney Brooks [13]. Brooks is well qualified for this analysis, as Director of the Artificial Intelligence Laboratory at MIT and also a chief officer in the iRobot Corporation.

A steady series of increasingly sophisticated robots have been designed specifically to interact with humans. Some have been laboratory devices (Kismet and Genghis). Others have been sold for the amusement of children (Furby, My Real Baby). And, primarily for the amusement of adults, there is AIBO, the Sony robot dog. iRobot Corporation sells a robotic home vacuum cleaner which is both fun and useful.

It is well known that the IBM chess playing computer Big Blue bested Gary Kasparov, the world's leading chess player. But, as noted by Brooks, Mr. Kasparov, in defeat, took satisfaction from the fact that Big Blue showed no evidence of elation in having won the match! So, clearly, Big Blue, lacking such an obvious emotion, is not human. Brooks[13] considers where this ongoing improvement may lead.

The sequence of robots mentioned does lead toward robots of human equivalent calculating power. Let's make the assumption that the AI/computer science community will eventually discover how to program such a supercomputing robot to make it willful and independent. (This idea is not new. Isaac Asimov's book, famously entitled "I, Robot", was written in 1950.)

In a Hollywood-like scenario, the intelligent robots become so much more advanced than the humans that they eventually decide to do away with the humans.

Brooks gives a set of hidden assumptions for such a (disastrous) outcome:

1. The machines can repair and reproduce themselves without human help.
2. It is possible to build machines that are intelligent but which do not have human emotions and, in particular, have no empathy for humans.
3. The machines we build will have a desire to survive and to control their environment to ensure their survival.
4. We, ultimately, will not be able to control our machines when they make a decision.

(End of quote from p. 200 of [13])

The status of these four assumptions is analyzed by Brooks, with a summary to the effect that one may expect many decades before any such disaster can occur.

It is also suggested, as by Fukuyama, that international rules of conduct may become appropriate (e.g., warlike aggressor robots are forbidden, not so different from the (presently void) ABM treaty).

To take a positive point of view, the benign areas of nanophysics and nanotechnology, are nonetheless likely to make contributions toward developments in the (possibly regulated, but doubtless active) areas of biotechnology and AI/robotics. In both medical and computer instrumentation, physics has been a strong source of innovation and insight. This is likely to accelerate, providing opportunities for nanotechnologists to be employed and to make contributions to advances.

8.5
Is there a Posthuman Future as Envisioned by Fukuyama?

The "Posthuman Future" [7] is a phrase to suggest a time period when the human condition will substantially change. The scenario particularly troubling to Fukuyama has to do with genetic engineering, and the possibility that humans may be subject to breeding not unlike breeding of dogs and other animals. This prospect has great opportunity and great danger. Roughly speaking, there has been success in genetically engineering new crops, new mice, and cloning of many animals has become routine. Why not engineer new humans! If one wants an athlete, there are various animals whose genes might make an addition! There is a lively debate between proponents and opponents of human intervention in what is called the "human germline".

One starting point is in the area of in-vitro fertilization, which has been beneficial to a small fraction of advanced human populations in overcoming infertility. The next step in this process is to perform more advanced screening on the genes of the several embryos that are often produced, and to offer the option of informed selection among these to promote an offspring with desired characteristics. Such screening is already used to exclude embryos that have undesired properties. (Of course, ultrasound screening of the normal in-vivo embryos has been used all over the world and has led in many places to very unnatural ratios of male to female births.) The opportunity seems to exist for an accelerated type of breeding of humans, not different in principle from the breeding of dogs or other animals.

There is further discussion of artificial genes being designed as a package that might be offered to augment any embryo, in the hope of achieving longer life or higher intelligence or other properties. The new genetic makeup would then propagate, and be perpetuated in offspring of such "super-human" or "post-human" beings. Some see this genetic engineering possibility as a potential breakthrough, to raise the capabilities of the human race, an opportunity not to be missed, a bit like populating outer space. Others see this as a chance for grave injustices, not to mention terrible mistakes. It seems it cannot occur without seriously violating the perceived rights of privacy and continuity of many, religious groups for example, probably constituting a majority. The issues are such that they may be easily misrepresented and easily misunderstood, or not understood at all. These are socially explosive issues.

The opportunity to breed a superior managerial class and an inferior laboring class would seem possible for a totalitarian regime, not to forget the attempt of the Nazi regime in prewar Germany to eradicate the Jewish population. Fukuyama [7] gives a careful discussion of this situation, and concludes that international controls of such developments are important. He quotes (p. 9) notably from Thomas Jefferson: "The general spread of the light of science has already laid open to every view the palpable truth, that the mass of mankind has *not* been born with saddles on their backs, nor a favored few booted and spurred, ready to ride them legitimately, *by the grace of God*" (Italics added). Most people disapprove of human slavery, which is at issue.

The analysis of Fukuyama [7] is restrained and statesmanlike. In contrast, the analysis of Brooks, in the area of germline engineering (changes in the human genome), is written more from an operational point of view, and imparts feelings of inevitability and urgency.

Quoting from Brooks [13] (pp. 235–236):

"It is clear that Robotic technology will merge with Biotechnology in the first half of this century…We are on a path to changing our genome in profound ways. Not simple improvements toward ideal humans as are often feared. In reality, we will have the power to manipulate our own bodies in the way we currently manipulate the design of machines. We will have the keys to our own existence. *There is no need to worry about mere robots taking over from us. We will be taking over from ourselves with manipulatable body plans and capabilities easily able to match that of any robot.* **The distinction between us and robots is going to disappear**" (Italics added).

So, one well informed answer to the question, "Will there be a posthuman future", is YES! Yes, our "posthuman future" seems very likely, and may well call for regulation!

This book has been about the benign and multidisciplinary science of nanophysics and the corresponding nanotechnology. These areas underlie the development of all futuristic scenarios, but they are not directly at the forefront (as we have demonstrated) of genetic engineering, artificial intelligence, or robotics. There is thus a large and ethically unencumbered opportunity for nanophysicsts and nanotechnologists to contribute, if so inclined, to "our posthuman future".

References

[1] From *Nanosystems: Molecular Machinery, Manufacturing & Computation* by K. Eric Drexler. Copyright © 1992 by Wiley Publishing, Inc. All rights reserved. Reproduced here by permission of the publisher.

[2] R. E. Smalley, Scientific American **285**, Number 3, 76 (2001).

[3] L. T. Scott, M. M. Boorum, B. J. McMahon, S. Hagen, M. Mack, J. Blank, H.Wagner, and A.de Meijere, Science **295**, 1500 (2002).

[4] R. C. Merkle, Nanotechnology **4**, 86 (1993).

[5] K. E. Drexler, *Engines of Creation*, (Anchor Books, New York, 1990) p. 172.

[6] Bill Joy, "Why the Future Does not Need us.", Wired magazine, April, 2000.

[7] F. Fukuyama, *Our Posthuman Future*, (Farrar, Straus and Giroux, New York, 2002) pp. 7, 8.

[8] Courtesy Cornelia Büchen-Osmond, Earth Institute, Columbia University.

[9] S. Albin, "Plague Patient Slowly Recovering", New York Times, January 17, 2003.

[10] D. Grady and G. Kolata, "Gene Therapy Used to Treat Patients with Parkinson's", New York Times, August 19, 2003.

[11] J. Cello, A. V. Paul, and E. Wimmer, Science **297**, 1016 (2002).

[12] R. Kurzweil, *The Age of Spiritual Machines* (Penguin, New York, 1999).

[13] R. A. Brooks, *Flesh and Machines: How Robots Will Change Us* (Vintage Books, New York, 2003) p. 200.

[14] R. A. Brooks, op.cit., pp. 235, 236.

Credits

IBM Research credit for Figures 3.9 and 7.3

Eastman Kodak Research for Figure 6.5

R. M. Eisberg and R. Resnick, *Quantum Physics of Atoms, Molecules, Solids, Nuclei, and Particles*, (2nd Edition) (Wiley, New York, 1985) for Figures 4.6, 5.1, 5.8, 5.9 and 5.10.

K. E. Drexler, *Nanosystems* (Wiley, New York, 1992) for Figures 5.3, 6.1 and 8.1

Charles Kittel, *Introduction to Solid State Physics* (Sixth Edition) (Wiley, New York, 1986), for Table 5.3, Figures 5.18 and 5.19

K. S. Krane, *Modern Physics* (2nd Edition) (Wiley, New York, 1995), for Figure 5.2.

F. J. Pilar, *Elementary Quantum Chemistry*, (Dover, New York., 2001), for Table 4.1 and Figures 4.7, 4.8 and 4.9.

Exercises

Exercises – Chapter 1

1. Referring to Figure 1.2: if there are 10 million transistors uniformly distributed on a one centimeter square silicon chip, what is the linear size of each unit?

2. A contemporary computer chip dissipates 54 Watts on an area of one centimeter square. Assuming that transistor elements in succeeding computer generations require constant power independent of their size (a hypothetical assumption), estimate the power that will be needed for a one centimeter square silicon chip in 5 years. Base your estimate on the Moore's Law trend of doubling the transistor count every 1.5 years. (The industry is confident of continuing this trend.)

3. Extrapolate the line in Figure 1.2, to estimate in which year the size of the transistor cell will be 10 nm.

4. In Figure 1.1, the vibrational motions of the single-crystal-silicon bars are transverse (vertical as seen in the picture) and the lowest frequency vibration corresponds to an anti-node at the middle, with nodes at each end of the bar. The resonances occur when $L = n\lambda/2$, $n = 1,2,3...$ If the fundamental frequency of the 2 micrometer (uppermost) bar is 0.4 GHz, what frequency does that bar have when oscillating in its 2nd harmonic? How many nodes will occur across the bar in that motion?

5. Regarding Figure 1.1, the supporting article states that the vibrations of the bars are generated with electromagnetic radiation. If the radiation used to excite the shortest (2 micrometer) bar is tuned to the bar's fundamental frequency of 0.4 GHz, what is the vacuum wavelength of the radiation?

6. Continuing from Exercise 5, suggest possible means of detecting the motion of a bar in order to confirm a resonance. [Hints: look in a transmission electron microscope (TEM), or Scanning Electron Microscope (SEM) for blurring of the image: look for power absorption from the source....]

Nanophysics and Nanotechnology: An Introduction to Modern Concepts in Nanoscience. Edward L. Wolf
Copyright © 2004 WILEY-VCH Verlag GmbH & Co. KGaA, Weinheim
ISBN: 3-527-40407-4

Exercises – Chapter 2

1. It is stated that the linear vibration frequency of the CO molecule is 64.2 THz. Using the masses of C and O as 12 and 16, respectively (in units of $u = 1.66 \times 10^{-27}$ kg), can you show that this vibration frequency is consistent with an effective spring constant K of 1860 N/m? [Hint: in a case like this the frequency becomes $\omega = 2\pi f = (K/\mu)1/2$, with effective mass $\mu = m_1 m_2/(m_1 + m_2)$.] What is the wavelength of radiation matching this frequency?

2. Treating the CO molecule of Exercise 1 as a classical oscillator in equilibrium with a temperature 300 K, estimate the amplitude of its thermal vibration.

3. Show that the spring constant K' of one piece of a spring cut in half is $K' = 2K$.

4. A small particle of radius 10 micrometers and density 2000 Kg/m^3 falls in air under the action of gravity. If the viscosity of air is 1.8×10^{-5} Pa·s, show that the terminal velocity is about 23 mm/s.

5. For the particle considered in Exercise 4, find the diffusion length during a time 1 s, in air, at 300 K.

Exercises – Chapter 3

1. In the Millikan "oil drop experiment", find the electric field in V/m needed to arrest the fall of a 10 micrometer radius oil particle with net charge $4e$. Take the density of the oil drop as 1000 Kg/m^3; the acceleration of gravity, g, as 9.8 m/s^2; and neglect the buoyant effect of the air in the chamber on the motion of the droplet.

2. An FM radio station transmits 1000 Watts at 96.3 MHz. How many photons per second does this correspond to? What is the wavelength? What is the significance of the wavelength, if any, from the photon point of view?

3. A photon's energy can be expressed as *pc*, where *p* is its momentum and *c* is the speed of light. Light carries energy and momentum but has zero mass! Calculate the force exerted by a 1000 Watt beam of light on a perfectly absorbing surface. Why is this force doubled if the surface, instead of absorbing, perfectly reflects all of the light?

4. It is found that short-wavelength light falling on a certain metal, causes emission (photoemission) of electrons of maximum kinetic energy, K, of 0.3 eV. If the work function, ϕ, of the metal in question is 4 eV, what is the wavelength of the light?

5. Explain, in the context of Exercise 4, how a value for Planck's constant h can be found as the slope of a plot of K vs. $f = c/\lambda$, where the light frequency f is assumed to be varied. Such experiments give the same value of Planck's constant h as was derived by Planck by fitting the spectrum of glowing light from a heated body. Einstein was awarded a Nobel Prize in part for his (1905) analysis of this *photoelectric effect*.

Exercises – Chapter 4

1. Using the Uncertainty Principle, estimate the minimum velocity of a bacterium (modeled as a cube of side 1 micrometer, and having the density of water), known to be located with uncertainty 0.1nm at $x = 0$, in vacuum and at $T = 0$.
2. Explain why the preceding question makes no sense if the bacterium is floating in water at 300 K. Explain, briefly, what is meant by Brownian motion.
3. Find the minimum speed of a C_{60} molecule in vacuum, if it is known to be located precisely at $x,y,z = 0$ plus or minus 0.01 nm in each direction.
4. A Buckyball C_{60} molecule of mass 1.195×10^{-24} Kg is confined to a one-dimensional box of length $L = 100$ nm. What is the energy of the $n = 1$ state? What quantum number n would be needed if the kinetic energy is 0.025 eV (appropriate to room temperature)?
5. A particle is in the $n = 7$ state of a one-dimensional infinite square well potential along the x-axis: $V = 0$ for $0 < x < L = 0.1$ nm; $V = $ infinity, elsewhere. What is the exact probability at $T = 0$ of finding the particle in the range 0.0143 nm $< x < 0.0286$ nm? (Answer is 1/7, by inspection of the plot of $P(x)$.)
6. Consider a hypothetical semiconductor with bandgap 1 eV, with relative electron mass 0.05, and relative hole mass 0.5. In a cube-shaped quantum dot of this material with $L = 3$ nm, find the energy of transition of an electron from the (211) electron state to the (111) hole state. (This transition energy must include the bandgap energy, and the additional energy is referred to as the "blue shift" of the fluorescent emission by the "quantum size effect".) Note that the "blue shift" can be tuned by adjusting the particle size, L.

Exercises – Chapter 5

1. Make an *estimate* of the vibrational frequency of the H_2 molecule, from the lower curve in Figure 5.1. (You can do this by approximating the minimum in the curve as $1/2\ K(r-r_o)^2$, where r is the interatomic spacing, and K is an effective spring constant, in eV/(Angstrom)^2. Estimate K from Figure 5.1; then adapt the usual formula for the frequency of a mass on a spring).
2. Use the result from Exercise 5.1 to find the "zero point energy" of di-hydrogen treated as a linear oscillator. Estimate the "zero-point motion" of this oscillator, in nm, and as a fraction of the equilibrium spacing. Compare this answer to one obtained using the uncertainty principle.
3. At $T = 300$ K, what is the approximate value of the oscillator quantum number, n, for the di-hydrogen vibrations, expected from $kT/2 = (n + 1/2)h\nu$? (At 300 K, an approximate value for kT is 0.025 eV.)
4. Explain, in a paragraph or two, the connections between the covalent bond energy of symmetric molecules, such as di-hydrogen; the notion of anti-symmetry under exchange for fermi particles, such as electrons; and the phenomenon of ferromagnetism (a basis for hard disk data storage).

5. Explain (a) why *mean free paths* of electrons and holes in semiconductors can be much larger than the spacing between the atoms in these materials. (b) Explain, in light of the nanophysical Kronig–Penney model, why the *mean free paths* of electrons and holes in exquisitely ordered solids (such as the semiconductor industry's 6 inch diameter single crystal "boules" of silicon) are not *infinitely* large (i.e., limited by the size, L, of the samples). Consider the effects of temperature, T, the impurity content N_i; and possible isotopic mass differences among the silicon atoms.

6. Why cannot a filled band provide electrical conductivity?

7. Why can the effective mass in a solid differ from the mass of the electron in vacuum?

8. Why does a hole have a charge of $+e$, and a mass equal to the mass of the valence band electron that moves into the vacant site?

9. How does Bragg reflection ($n\lambda = d\sin\theta$) influence the motion of carriers in a (one-dimensional) semiconductor with lattice constant a? Why does this lead to the answer that the "zone-boundary" is at $k = \pi/a$?

10. Explain why the group velocity of carriers is zero at "the zone boundary", $k = \pi/a$.

11. Explain the difference between the two (standing wave) wavefunctions valid at $k = \pi/a$. Explain, qualitatively, the difference in their electrostatic energies, and why this energy difference is precisely the "band gap" into the next higher band.

Exercises – Chapter 6

1. The basic measure of thermal agitation energy is $1/2\,kT$ per degree of freedom, where k is Boltzmann's constant. How does thermal energy (at 400 K) compare with the energy differences among the computer-modeled simulations of octane presented in Figure 6.1?

2. In Figure 6.1, 10^{11} Hz is a characteristic rate at which the octane molecule switches from one shape (conformation) to another. For comparison, find the lowest longitudinal vibration frequency, using the methods of Chapter 2, modeling octane as a chain of eight carbon atoms (mass 12) connected by springs of strength $K = 440$ N/m and length 0.15 nm.

3. The variety of different conformations of octane, exhibited in Figure 6.1, seem incompatible with using octane to carry a single (small-diameter) tip, which is a required function of the proposed "molecular assembler". With regard to stability, give reasons why acetylene C_2H_2 (a) and a small-diameter carbon nanotube (b), should be superior.

4. Consider the FET device shown in Figure 6.4. Why is the forward current in this FET so large? How could such FET devices be assembled at a density of 10^7/cm^2 on a silicon chip? (If you have a good answer, have your statement notarized and hire a patent lawyer!)

5. How fast can the FET device shown in Figure 6.4 respond to a gate voltage turned on at time zero? It is said that the electron transport along a nanotube is "ballistic", which means that, if the force is eE, then the velocity is eEt/m, and the distance covered in time t is $1/2(eE/m)t^2$. For the device geometry shown, where $E = V/L$, set $V = 1$ Volt and find: (a) the corresponding value of t. Secondly, (b) for the device geometry shown, convert this result into an effective mobility μ, such that $v = \mu E = \mu V/L$, and $t = L^2/\mu V$. Note that the mobility μ is conventionally given in units of cm^2/Vs. Compare this mobility to values listed in Table 5.1.

6. Consider the magnetotactic bacterium shown in Figure 6.6. (a) Calculate the torque exerted on the bacterium in an earth's magnetic field of 1 Gauss (10^{-4} T) if the bacterium is oriented at 90°. (b) Explain why a strain of magnetotactic bacteria found in the northern hemisphere, if transported to an ocean in the southern hemisphere, would surely die out.

7. In the STM study of ordered arrays of C_{60} shown in Figure 6.8 it was observed that the molecules were locked into identical orientations at 5 K, but were apparently rotating at 77 K. (a) From this information, make an estimate of the interaction energy between nearest-neighbor C_{60} molecules. Is this energy in the right range to be a van der Waals interaction? (b) Compare this temperature information to the melting point of a "C_{60} molecular solid" which you can estimate from Figure 5.2.

Exercises – Chapter 7

1. Compare the conventional, state-of-the-art, spatial resolution for modern photolithography, presented as 180 nm, to the resolution, in principle, that is available at wavelength 248 nm, see Figure 7.3.

2. Explain why e-beam writing is not competitive as a production process for writing wiring on Pentium chips.

3. What features of Figure 7.3 are inconsistent with the stated exposure wavelength of 248 nm?

4. Compare e-beam writing to conventional lithography with regard to *resolution* and *throughput*.

5. It is reported in Consumer Reports (October 2003, p. 8) that the "Segway Human Transporter" has five independent *electronic gyroscope sensors*, with, presumably, a corresponding set of servo systems. *Write an essay on how such rotation sensors might be fabricated (clue, one method might be based on Figure 7.1).* In principle, how would the performance of such rotation sensors be expected to change with a uniform reduction in scale, *xyz* of × 10?

6. What are the chief advantages of the superconductive RSFQ computing technology, and what are the chief disadvantages of the same technology?

7. Why is the spatial resolution of the STM better than that of the AFM?

8. Why can't a single tip in the STM/AFM technology possibly perform operations at a rate greater then 1 GHz?

9. Why cannot a single tip in the STM/AFM technology perform operations beyond *observing*, *nudging*, and *exciting* an atom (or molecule) in its view? (Explain what is meant by "nudging", and why it is easier for atoms, like Xe, or molecules, like benzene, with a large number of electrons.) Other desirable, but unavailable, functions for a single tip would be picking up, orienting, and depositing, atoms or molecules. *Why are these functions unavailable?*

Exercises – Chapter 8

1. Estimate the diameters of the proposed molecules in Figure 8.1.
2. Provide a critique of the suggested size of the "molecular assembler" as one billion atoms (it may be on the small side?) What would be the expected radius for a spherical assembly of one billion atoms, in nm? Can you estimate the number of atoms in a typical enzyme, as a comparison?
3. It has been suggested that the time for a typical bacterium to grow is about an hour. Suppose a rectangular (cube) bacterium has dimension $L = 1$ micrometer, the density of water, and is made of carbon 12. Roughly how many atoms are there in this bacterium? Calculate the rate of growth in atoms per second if this forms in one hour. Compare your answer to the performance of the IBM Millipede AFM, discussed in Chapter 7.
4. The example of the magnetotactic bacterium (see Chapter 6) proves that some robust inorganic solids, eg. Fe_3O_4, can be grown at room temperature from (ionic) solution. Compare this case with the case of diamond (and the proposed diamondoid inorganic molecules), which require high temperatures/pressures to form (and apparently cannot be grown from solution). What is the crucial difference?
5. What is the most significant difference between a virus and a living organism? (Can a virus live forever?)
6. Poll your colleagues to ask if they think that "computers" (or artificially intelligent robots) will ever be able to establish separate identities, consciousness, and other human qualities. If so, how threatening does this prospect seem?

Index

Nanophysics and Nanotechnology: An Introduction to Modern Concepts in Nanoscience. Edward L. Wolf
Copyright © 2004 WILEY-VCH Verlag GmbH & Co. KGaA, Weinheim
ISBN: 3-527-40407-4

Nanotechnology

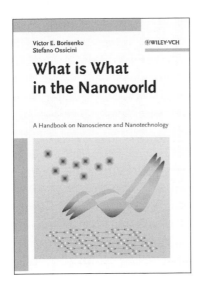

VICTOR E. BORISENKO, Belarusian
State University, and STEFANO
OSSICINI, University of Modena and
Reggio Emilia

What is What the in Nanoworld
A Handbook on Nanoscience and Nanotechnology

2004. XII, 335 pages, 120 figures, 28
tables. Hardcover.

ISBN 3-527-40493-7

This introductory, reference handbook summarizes the terms and definitions, most important phenomena, and regulations discovered in the physics, chemistry, technology, and application of nano-structures. The short form of information taken from textbooks, special encyclopedias, recent original books and papers provides fast support in understanding "old" and new terms of nanoscience and technology widely used in scientific literature on recent developments. Additional information in the form of notes supplements entries and gives a historical retrospective of the subject with reference to further sources.
Ideal for answering questions related to unknown terms and definitions of undergraduate and Ph.D. students when studying the physics of low-dimensional structures, nanoelectronics, nanotech-nology.

Wiley-VCH
P.O. Box 10 11 61 • D-69451 Weinheim, Germany
Fax: +49 (0)6201 606 184
e-mail: service@wiley-vch.de • www.wiley-vch.de

Nanotechnology

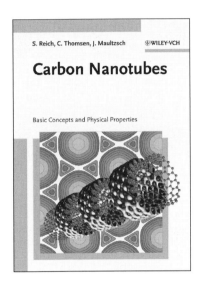

STEPHANIE REICH, University of Cambridge, UK

CHRISTIAN THOMSEN, Technical University of Berlin, Germany

JANINA MAULTZSCH, Technical University of Berlin, Germany

Carbon Nanotubes
Basic Concepts and Physical Properties

2004. IX, 215 pages, 126 figures. Hardcover.
ISBN 3-527-40386-8

This text is an introduction to the physical concepts needed for investigating carbon nanotubes and other one-dimensional solid-state systems. Written for a wide scientific readership, each chapter consists of an instructive approach to the topic and sustainable ideas for solutions. The former is generally comprehensible for physicists and chemists, while the latter enable the reader to work towards the state of the art in that area. The book gives for the first time a combined theoretical and experimental description of topics like luminescence of carbon nanotubes, Raman scattering, or transport measurements. The theoretical concepts discussed range from the tight-binding approximation, which can be followed by pencil and paper, to first-principles simulations. The authors emphasize a comprehensive theoretical and experimental understanding of carbon nanotubes.

Wiley-VCH
P.O. Box 10 11 61 • D-69451 Weinheim, Germany
Fax: +49 (0)6201 606 184
e-mail: service@wiley-vch.de • www.wiley-vch.de